Pseudo-Differential Operators
Theory and Applications
Vol. 9

Managing Editor

M.W. Wong (York University, Canada)

Editorial Board

Luigi Rodino (Università di Torino, Italy)
Bert-Wolfgang Schulze (Universität Potsdam, Germany)
Johannes Sjöstrand (Université de Bourgogne, Dijon, France)
Sundaram Thangavelu (Indian Institute of Science at Bangalore, India)
Maciej Zworski (University of California at Berkeley, USA)

Pseudo-Differential Operators: Theory and Applications is a series of moderately priced graduate-level textbooks and monographs appealing to students and experts alike. Pseudo-differential operators are understood in a very broad sense and include such topics as harmonic analysis, PDE, geometry, mathematical physics, microlocal analysis, time-frequency analysis, imaging and computations. Modern trends and novel applications in mathematics, natural sciences, medicine, scientific computing, and engineering are highlighted.

Leon Cohen

The Weyl Operator
and its Generalization

Leon Cohen
Hunter College & Graduate Center
City University of New York
New York
USA

ISBN 978-3-0348-0293-2   ISBN 978-3-0348-0294-9 (eBook)
DOI 10.1007/978-3-0348-0294-9
Springer Basel Heidelberg New York Dordrecht London

Library of Congress Control Number: 2012955049

Mathematics Subject Classification (2010): 35S05, 53D55, 81Sxx, 81S30, 81S05, 46L65

© Springer Basel 2013
This work is subject to copyright. All rights are reserved by the Publisher, whether the whole or part of the material is concerned, specifically the rights of translation, reprinting, reuse of illustrations, recitation, broadcasting, reproduction on microfilms or in any other physical way, and transmission or information storage and retrieval, electronic adaptation, computer software, or by similar or dissimilar methodology now known or hereafter developed. Exempted from this legal reservation are brief excerpts in connection with reviews or scholarly analysis or material supplied specifically for the purpose of being entered and executed on a computer system, for exclusive use by the purchaser of the work. Duplication of this publication or parts thereof is permitted only under the provisions of the Copyright Law of the Publisher's location, in its current version, and permission for use must always be obtained from Springer. Permissions for use may be obtained through RightsLink at the Copyright Clearance Center. Violations are liable to prosecution under the respective Copyright Law.
The use of general descriptive names, registered names, trademarks, service marks, etc. in this publication does not imply, even in the absence of a specific statement, that such names are exempt from the relevant protective laws and regulations and therefore free for general use.
While the advice and information in this book are believed to be true and accurate at the date of publication, neither the authors nor the editors nor the publisher can accept any legal responsibility for any errors or omissions that may be made. The publisher makes no warranty, express or implied, with respect to the material contained herein.

Printed on acid-free paper

Springer Basel is part of Springer Science+Business Media (www.springer.com)

*To Carol, Valerie, Ken, Livia,
Douglas, Rummy, and Polar*

Contents

Preface   xi

1 Introduction and Terminology   1
   1.1 The Fundamental Issue 2
   1.2 Notation and Terminology in Different Fields 3

2 Operator Algebra   5
   2.1 Exponential Operator 8
   2.2 Manipulating $D^m X^n$ and $X^m D^n$ 8
   2.3 Translation Operator 9
   2.4 The Operator $e^{i\theta X + i\tau D}$ 10
   2.5 The Operator $e^{\xi H} A e^{-\xi H}$ 12
   2.6 Phase-Space Operator Formulas 13
   2.7 Transforms and the Representation of Functions 14
   2.8 Delta Function 17
   2.9 Intensity, Probability, and Averages 19
   2.10 Simplifying Functions of Operators 20
   2.11 Rearrangement of Operators 22

3 The Weyl Operator   25
   3.1 The Weyl Correspondence 25
   3.2 Operation on a Function 26
   3.3 Operation in the Fourier Domain 27
   3.4 Operational Form 27
   3.5 Examples 28
   3.6 Inversion: From Weyl Operator to Symbol 32
   3.7 Adjoint 36
   3.8 Hermiticity 36
   3.9 The Algebra of the Weyl Operator 37
   3.10 The Wigner distribution 41
   3.11 Product of Weyl Operators, Commutators, Moyal Bracket 43
   3.12 Operator Form 44

4 Generalized Operator Association — 47
- 4.1 Generalized Operator — 47
- 4.2 Operational Form — 48
- 4.3 The Operation on a Function — 49
- 4.4 From Operator to Symbol — 50
- 4.5 Kernel From a Monomial Rule — 50
- 4.6 Algebra — 52
- 4.7 Hermitian Adjoint — 53
- 4.8 Product of Operators — 54
- 4.9 Transformation Between Associations — 55
- 4.10 The Fourier, Taylor, and Delta Function Associations — 56
- 4.11 The Form of the Generalized Correspondence — 58
- 4.12 The Kernel for the Born-Jordan Rule — 58

5 Generalized Phase-Space Distributions — 61
- 5.1 Characteristic Function Approach — 63
- 5.2 Marginal Conditions — 64
- 5.3 Relation Between Distributions — 64
- 5.4 Manifestly Positive Distributions — 66
- 5.5 Singular Kernels — 67

6 Special Cases — 69
- 6.1 Summary — 69
- 6.2 Standard Ordering — 70
- 6.3 Anti-Standard Ordering — 72
- 6.4 Symmetrization Ordering — 73
- 6.5 Born-Jordan Ordering — 73
- 6.6 Choi-Williams Ordering — 74
- 6.7 Weyl Ordering — 75
- 6.8 ZAM Ordering — 76
- 6.9 Spectrogram — 78
- 6.10 Gaussian Window — 80
- 6.11 One Parameter Families — 81
- 6.12 Normal and Anti-Normal Ordering — 82

7 Unitary Transformation — 85
- 7.1 Unitary and Hermitian Operators — 86
- 7.2 Unitary Transformation of the Generalized Association — 87
- 7.3 Transformation of the Generalized Distribution — 88
- 7.4 Examples — 89

8 Path Integral Approach — 91
- 8.1 Configuration Space — 91
- 8.2 Phase-Space — 93

9 Time-Frequency Operators — 95
- 9.1 Time-Frequency Association Rules 96
- 9.2 Time-Frequency Distributions 98
- 9.3 Complex Signals and Instantaneous Frequency 98
- 9.4 Time-Frequency Space-(Spatial) Frequency 100

10 Transformation of Differential Equations Into Phase Space — 103
- 10.1 Transformation Properties of the Wigner Distribution 103
- 10.2 Ordinary Differential Equations 105
- 10.3 Non-Constant Coefficients 106

11 The Eigenvalue Problem in Phase-Space — 107
- 11.1 General Kernel 109

12 Arbitrary Operators: Single Operator — 111
- 12.1 The Probability Distribution Corresponding to an Operator 111
- 12.2 Expectation Value 114
- 12.3 Examples 114

13 Uncertainty Principle for Arbitrary Operators — 121
- 13.1 The Standard Deviation of an Operator 121
- 13.2 Schwarz Inequality 123
- 13.3 Uncertainty Principle 124
- 13.4 Examples 125

14 The Khintchine Theorem and Characteristic Function Representability — 129

15 Arbitrary operators: Two Operators — 131
- 15.1 Operator Association 131
- 15.2 Joint Distribution 132
- 15.3 Commuting Operators 133
- 15.4 The $e^{i\theta A + i\tau B}$ Correspondence 135
  - 15.4.1 Disentanglement of $e^{i\theta A + i\tau B}$ 136
  - 15.4.2 General procedure for evaluating $e^{i\theta A + i\tau B} \varphi(x)$ 137
  - 15.4.3 Linear Combination of X and D 138
- 15.5 The $e^{i\theta A} e^{i\tau B}$ Correspondence 146
- 15.6 The $e^{i\tau A/2} e^{i\theta B} e^{i\tau A/2}$ Correspondence 148

Bibliography — 151

Index — 157

Preface

The concept of associating ordinary functions with operators has arisen in many areas of science and mathematics and it can be argued that the earliest instance was Leibniz's attempt to define a fractional derivative. Up to the beginning of the twentieth century, many isolated results were obtained and culminated with the remarkable contributions of Heaviside and the efforts to put his methods on a sound mathematical footing. These developments were mostly based on associating a function of one variable with one operator, the operator generally being the differentiation operator.

With the discovery of quantum mechanics in the years 1925-1930, there arose, in a natural way, the issue that one has to associate a function of two variables with a function of two operators that do not commute. This has led to a wonderfully rich mathematical development that has found applications in many fields, including pseudo-differential operators, time-frequency analysis, quantum optics, wave propagation, differential equations, image processing, radar, sonar, chemical physics, and acoustics, among others. The earliest proposal for associating an ordinary function of two variables with an operator was that of Born and Jordan (1925), and subsequently Weyl (1929) and others proposed other rules. There are an infinite number of ways to associate a function of two ordinary variables with a function of two operators because ordinary variables commute while operators generally do not. The rules became known as rules of association, correspondence rules, or ordering rules.

Independently of these developments Wigner, in 1932, and Kirkwood, in 1933 devised a classical-like joint distribution where one can calculate operator averages in the standard probability manner, that is, by phase-space integration. No connection was made between these distributions and correspondence rules until Moyal in 1949 saw things clearly. He *derived* the Wigner distribution using the Weyl correspondence. Subsequently it was realized that for every correspondence rule there is a corresponding phase-space distribution. Now the field of correspondence rules and phase-space distributions are intimately connected. Remarkably, around the same time as Moyal, a similar development occurred in the field of time-varying spectral analysis whose aim was to understand signals with changing frequencies, human speech being the prime historical example. It was realized by Gabor and Ville that one can define time and frequency operators and make a mathematical analogy with the quantum case.

I have aimed to present the basic ideas and results of correspondence rules in a straightforward elementary manner and have included many examples to illustrate the ideas developed. I have strived to make the mathematics accessible to a wide audience and I have avoided delving into advanced formulations such as group theoretical considerations. The level of rigor and terminology are those of the original contributors and standard in mathematical physics and engineering.

I would like to thank Man Wah Wong for his friendship and for encouraging me to write the book. I express my deep appreciation to Livia Cohen, Lorenzo Galleani, Barnabas Kim, Patrick Loughlin, and Jeruz Tekel for reading the manuscript and for their many valuable suggestions. I thank the Office of Naval Research for support of my research.

Chapter 1

Introduction and Terminology

This book deals with correspondence rules or rules of association. The fundamental idea is to associate a function of ordinary variables with an operator. While in later chapters we deal with arbitrary operators, in the main portion of the book we deal with the operators X and D where,

$$X = \begin{cases} x & \text{in the } x \text{ representation} \\ i\frac{d}{dp} & \text{in the Fourier representation,} \end{cases} \quad (1.1)$$

$$D = \begin{cases} \frac{1}{i}\frac{d}{dx} & \text{in the } x \text{ representation} \\ p & \text{in the Fourier representation.} \end{cases} \quad (1.2)$$

The fundamental relation between X and D is the commutator

$$[X, D] = XD - DX = i. \quad (1.3)$$

Depending on the field, these operators may be appropriately called position and spatial frequency, or position and momentum, and in time-frequency analysis they correspond to the time and frequency operators.

One of the basic reasons for considering these particular operators is that we can use them to evaluate expectation values of functions in the Fourier or x representation without leaving the representation. In particular, suppose we have a function, $f(x)$, and a "state" function $\varphi(x)$; then

$$\int f(x)\,|\varphi(x)|^2\,dx = \int \widehat{\varphi}^*(p)f(X)\widehat{\varphi}(p)\,dp \quad (1.4)$$

where $\widehat{\varphi}(p)$ is the Fourier transform of $\varphi(x)$,

$$\widehat{\varphi}(p) = \frac{1}{\sqrt{2\pi}} \int \varphi(x)\,e^{-ixp}\,dx. \quad (1.5)$$

Therefore, we say that $f(x)$ in the x domain is associated with or represented by the operator $f(X)$ in the Fourier domain and we write

$$f(X) \leftrightarrow f(x) \tag{1.6}$$

where $\leftrightarrow$ indicates the association. Furthermore, suppose we have a function of p, $g(p)$, in the Fourier domain, then

$$\int g(p) \, |\widehat{\varphi}(p)|^2 \, dp = \int \varphi^*(x) g(D) \varphi(x) \, dx \tag{1.7}$$

and we say that $g(p)$ in the Fourier domain is associated with $g(D)$ in the x domain,

$$g(D) \leftrightarrow g(p). \tag{1.8}$$

The right hand side of Eq. (1.7) is sometimes called *sandwiching* because the operator, $g(D)$, is in between $\varphi^*(x)$ and $\varphi(x)$.

The basic advantage of Eq. (1.7) is that if we want to calculate the expectation of $g(p)$ as defined by the left hand side we do not have to first calculate $\widehat{\varphi}(p)$. We can remain in the x representation and use the right hand side to calculate it.

1.1 The Fundamental Issue

In the above discussion we had no difficulty in associating a function of one variable $f(x)$ or $g(p)$ with its corresponding operator. But what if we have a function of two variables, for example xp^2, then what will the association be? It could be, for example, XD^2, DXD, or D^2X, among others; all of these associations are proper in the sense that they reduce to xp^2 if we just think of the operators as ordinary variables. However, all these choices are different because X and D do not commute. Formulating such associations for a general function $a(x,p)$ is the fundamental aim. There have been many rules proposed, among them the Weyl, Standard, and Born-Jordan. In the next chapter we study the Weyl rule and subsequently other rules. In Chap. 4 we present a general method that handles all rules in a unified manner.

Furthermore, suppose we did have an association for the function $a(x,p)$ with an operator $A(X,D)$. Then, what is the generalization of Eqs. (1.4) and (1.7)? We will see that the proper generalization is that for a state function $\varphi(x)$ we have to introduce a joint function, $C(x,p)$ called the generalized distribution function which will allow us to write

$$\int \varphi^*(x) A(X,D) \varphi(x) \, dx = \iint a(x,p) C(x,p) \, dx \, dp \tag{1.9}$$

As we will see in Chap. 5 this is the generalization of Eqs. (1.4) and (1.7) and reduces to them when the symbol is a function of x or p only.

1.2 Notation and Terminology in Different Fields

Generally speaking there are five fields where these methods have been used. They are quantum mechanics, pseudo-differential operators, time-frequency analysis, wave propagation, and image processing. We review here some mathematical conventions and contrast the differences in terminology used in various fields.

"Symbol", "classical function", "c-function", and *"ordinary function"* are terms used in different fields to signify the same thing, namely an ordinary function, $a(x,\xi)$, of two variables x and ξ. This is the common notation used in mathematics, while in quantum mechanics it is position and momentum signified by $a(q,p)$, and in time-frequency analysis one generally writes $a(t,\omega)$. In the case of wave propagation and image processing one usually uses $a(x,k)$ where x and k are respectively position and wave-number (spatial-frequency). We will use (x,p).

"Association", "correspondence rule," "rule of association," "quantization," and *"ordering rule."* These words and phrases all mean the same thing, namely the association of an operator $A(X,D)$ with an ordinary function, $a(x,p)$. The two variables, x and p, are generally conjugate variables. Conjugate generally means that their corresponding operators do not commute. For the Weyl case, the association is symbolized by

$$A(X,D) \leftrightarrow a(x,p). \tag{1.10}$$

We call $A(X,D)$ the Weyl operator. When we study other correspondences we use

$$A^{\mathrm{W}}(X,D) \leftrightarrow a(x,p) \tag{1.11}$$

where the superscript denotes the particular correspondence, in this case the Weyl rule. Similarly for other cases, as, for example

$$A^{\mathrm{BJ}}(X,D) \leftrightarrow a(x,p) \tag{1.12}$$

will denote the Born-Jordan correspondence.

The state function, signal, and image. Generally speaking the state function is an ordinary function upon which the operator operates. In engineering it is called the *signal* and in image processing it is called the image. In acoustics it is the pressure wave and in electrodynamics it may be the electric or magnetic field. The term state function comes from quantum mechanics where it represents the state of the physical system at hand. This is in contrast to the operators which represent physical observables. The state function may be square integrable or not and can be a distribution in the delta function sense.

Phase-Space and quasi-distributions. By a phase-space function we mean a function of the two variables x and p. For example the symbol $a(x,p)$ is a phase-space function. The Wigner and other distributions are also phase-space functions and are called distributions. The terminology "distribution" comes from usage in physics and chemistry and are densities in probability theory. The phase-space distributions we consider in this book are generally not manifestly positive and that is why they are sometimes called quasi-distributions, but we will simply use the term distribution.

Fourier transform pairs. The Fourier transform of the symbol, $a(x,p)$, will be denoted by $\hat{a}(\theta,\tau)$ and the normalization is taken so that

$$\hat{a}(\theta,\tau) = \frac{1}{4\pi^2} \iint a(x,p)\, e^{-i\theta x - i\tau p}\, dx\, dp, \qquad (1.13)$$

$$a(x,p) = \iint \hat{a}(\theta,\tau)\, e^{i\theta x + i\tau p}\, d\theta\, d\tau. \qquad (1.14)$$

However, due to conventions in different fields we define Fourier transform pairs for the state function by

$$\hat{\varphi}(p) = \frac{1}{\sqrt{2\pi}} \int \varphi(x)\, e^{-ixp}\, dx, \qquad (1.15)$$

$$\varphi(x) = \frac{1}{\sqrt{2\pi}} \int \hat{\varphi}(p)\, e^{ixp}\, dp, \qquad (1.16)$$

and similarly for higher dimensions. The advantage of this convention is that it keeps the symmetry between the two domains in the sense that

$$\int |\varphi(x)|^2\, dx = \int |\hat{\varphi}(p)|^2\, dp \qquad (1.17)$$

and also keeps the symmetry in Eq. (1.4) and (1.7).

Commutator and anti-commutator. The commutator of two operators A and B is standardly denoted by

$$[A,B] = AB - BA \qquad (1.18)$$

and the anti-commutator by

$$[A,B]_+ = AB + BA. \qquad (1.19)$$

The D operator. We will occasionally use $D_y = \frac{1}{i}\frac{d}{dy}$, etc., otherwise D is understood to be D_x.

Integrals. Integrals without limits imply integration over the reals,

$$\int = \int_\mathbb{R} = \int_{-\infty}^{\infty}. \qquad (1.20)$$

Chapter 2

Operator Algebra

An operator transforms one function into another. One writes and says that A operates on f to produce g. We generally assume that operators are linear which means that for any two functions, f_1 and f_2,

$$A(f_1 + f_2) = Af_1 + Af_2. \tag{2.1}$$

When we say that two operators are equal

$$A = B \tag{2.2}$$

one means that for an arbitrary function, $f(x)$,

$$Af = Bf. \tag{2.3}$$

For example, the two operators D^2XD and $XD^3 - 2iD^2$ are equal, because indeed when they operate on an arbitrary function they give the same answer. Specifically

$$D^2xDf(x) = (xD^3 - 2iD^2)f(x). \tag{2.4}$$

If we wanted to verify this we could write it explicitly in terms of derivatives,

$$\frac{1}{i^3}\frac{d^2}{dx^2}x\frac{d}{dx}f(x) = \left(\frac{1}{i^3}x\frac{d^3}{dx^3} - 2i\frac{1}{i^2}\frac{d^2}{dx^2}\right)f(x) \tag{2.5}$$

and carry out the differentiation of both sides to verify the identity. One can also do it in the Fourier domain and verify that indeed

$$p^2 Xp\,\widehat{f}(p) = (Xp^3 - 2ip^2)\,\widehat{f}(p) \tag{2.6}$$

or explicitly

$$p^2 i\frac{d}{dp}p\,\widehat{f}(p) = \left(i\frac{d}{dp}p^3 - 2ip^2\right)\widehat{f}(p). \tag{2.7}$$

However, a much better way is to manipulate the operators algebraically by using the commutation relation, $DX = XD - i$. In particular

$$D^2XD = DDXD = D(XD - i)D = DXD^2 - iD^2 \qquad (2.8)$$
$$= (XD - i)D^2 - iD^2 = XD^3 - 2iD^2. \qquad (2.9)$$

Of course, in manipulating operator equations, attention must be paid to the fact that operators generally do not commute.

We now give some basic definitions and results regarding operators.

Inverse. The inverse of an operator, A^{-1}, is an operator such that

$$A^{-1}A = AA^{-1} = I \qquad (2.10)$$

where I is the identity operator. The inverse of the product of two operators is given by

$$(AB)^{-1} = B^{-1}A^{-1}. \qquad (2.11)$$

Adjoint. The adjoint of an operator is another operator, denoted by $A^\dagger$, which forces the equality

$$\int g^* A f \, dx = \int f \{A^\dagger g\}^* \, dx \qquad (2.12)$$

where g and f are any two functions. The adjoint of a product of operators is given by

$$(AB)^\dagger = B^\dagger A^\dagger. \qquad (2.13)$$

The adjoint is sometimes called the Hermitian adjoint.

Hermitian operator. If the adjoint of an operator equals the operator

$$A = A^\dagger, \qquad (2.14)$$

one then says that the operator is a self adjoint or a Hermitian operator. For a Hermitian operator Eq. (2.12) becomes

$$\int g^* A f \, dx = \int f \{Ag\}^* dx \qquad (2.15)$$

which may be taken as the definition of a Hermitian operator. The basic operators X and D are Hermitian operators. Hermitian operators play a particularly important role for many reasons that will be discussed in subsequent sections, but the most important is the eigenvalues of a Hermitian operator are real and the eigenfunctions are orthogonal and complete.

Chapter 2. Operator Algebra

Expressing an arbitrary operator in terms of Hermitian operators. An arbitrary operator, A, can be written as

$$A = \tfrac{1}{2}(A+A^\dagger) + \tfrac{1}{2}i(A-A^\dagger)/i. \tag{2.16}$$

One can readily prove that both $(A+A^\dagger)$ and $(A-A^\dagger)/i$ are Hermitian. Eq. (2.16) shows that we can express any operator as a sum of a Hermitian operator plus i times a Hermitian operator. This is the operator analog of writing a complex number in terms of its real and imaginary parts.

Product of two operators. The product of two Hermitian operators is generally not Hermitian. However, writing AB as

$$AB = \tfrac{1}{2}[A,B]_+ + \tfrac{i}{2}[A,B]/i \tag{2.17}$$

expresses AB in terms of the commutator and anti-commutator. Moreover $[A,B]_+$ and $[A,B]/i$ are Hermitian.

Unitary Operator. An operator U is said to be unitary if its adjoint is equal to its inverse,

$$U^\dagger = U^{-1}. \tag{2.18}$$

Unitary operators are discussed in Chap. 7 but we mention here that the importance of being unitary is that when it operates on a function, normalization is preserved in the sense that

$$\int |f(x)|^2 dx = \int |Uf(x)|^2 dx. \tag{2.19}$$

An operator of the form

$$U = e^{iA} \tag{2.20}$$

is unitary if A is Hermitian and conversely. This is sometimes known as Stone's theorem.

Functions of operators. There are many ways to define functions of operators and we mention two of them. One way is via a power series. A function $f(A)$ of the operator A shall mean that we expand the ordinary function $f(x)$ in a power series

$$f(x) = \sum_{n=0}^{\infty} f_n x^n \tag{2.21}$$

and then substitute A for x,

$$f(A) = \sum_{n=0}^{\infty} f_n A^n. \tag{2.22}$$

A second way is by way of the Fourier transform of $f(x)$. Define

$$\widehat{f}(p) = \frac{1}{\sqrt{2\pi}} \int f(x) e^{-ixp} dx, \qquad (2.23)$$

we then define $f(A)$ by

$$f(A) = \frac{1}{\sqrt{2\pi}} \int \widehat{f}(p) e^{iAp} dp. \qquad (2.24)$$

2.1 Exponential Operator

The exponential operator, e^A, where A is an operator, plays a fundamental role in correspondence rules and in many other branches of science and mathematics [95]. We discuss here some of its basic properties and further developments will be discussed as needed.

Differentiation of $e^{A(\theta)}$. If $A(\theta)$ is an operator that depends on a parameter θ, then

$$\frac{d}{d\theta} e^{A(\theta)} = \int_0^1 e^{(1-s)A(\theta)} \frac{dA}{d\theta} e^{sA(\theta)} ds = \int_0^1 e^{sA(\theta)} \frac{dA}{d\theta} e^{(1-s)A(\theta)} ds. \qquad (2.25)$$

A special case that often arises is the differentiation of $e^{A+\theta B}$ with respect to θ but where A and B are operators that do not depend on θ. In this case

$$\frac{d}{d\theta} e^{A+\theta B} = \int_0^1 e^{(1-s)(A+\theta B)} B e^{s(A+\theta B)} ds. \qquad (2.26)$$

Also of interest is the evaluation of Eq. (2.26) at zero,

$$\frac{d}{d\theta} e^{A+\theta B} \bigg|_{\theta=0} = \int_0^1 e^{(1-s)A} B e^{sA} ds. \qquad (2.27)$$

2.2 Manipulating $D^m X^n$ and $X^m D^n$

In manipulating expressions involving X and D the following relations are useful,

$$D^m X^n = \sum_{k=0}^{\min(m,n)} (-i)^k k! \binom{n}{k} \binom{m}{k} X^{n-k} D^{m-k} \qquad (2.28)$$

and

$$X^n D^m = \sum_{k=0}^{\min(m,n)} i^k k! \binom{n}{k} \binom{m}{k} D^{m-k} X^{n-k}. \qquad (2.29)$$

2.3. Translation Operator

These expressions were first derived by McCoy [55]. A few other important relations are

$$[X, D^n] = inD^{n-1} \tag{2.30}$$
$$[X^n, D] = inX^{n-1} \tag{2.31}$$
$$[D, f(x)] = \left(\frac{1}{i}\frac{d}{dx}f(x)\right) = (Df(x)). \tag{2.32}$$

In Eq. (2.32) we have put parentheses around $\frac{1}{i}\frac{d}{dx}f(x)$ to emphasize that the differentiation is only on $f(x)$ and not on the function that $[D, f(x)]$ operates on. That is, for an arbitrary function $g(x)$,

$$[D, f(x)]g(x) = \left(\frac{1}{i}\frac{d}{dx}f(x)\right)g(x). \tag{2.33}$$

2.3 Translation Operator

The operator $e^{i\tau D}$ is the translation operator in the x representation

$$e^{i\tau D}f(x) = f(x+\tau). \tag{2.34}$$

This is readily proven. Since

$$e^{i\tau D}f(x) = \sum_{n=0}^{\infty} \frac{(i\tau)^n D^n}{n!}f(x) = \sum_{n=0}^{\infty} \frac{\tau^n}{n!}\frac{d^n}{dx^n}f(x) \tag{2.35}$$

which is precisely the Taylor expansion of $f(x+\tau)$ and therefore Eq. (2.34) follows. Since D is Hermitian, τD is Hermitian for real τ, the translation operator is hence unitary. For translation in the Fourier domain we have

$$e^{-i\theta X}\widehat{\varphi}(p) = e^{\theta \frac{\partial}{\partial p}}\widehat{\varphi}(p) = \widehat{\varphi}(p+\theta). \tag{2.36}$$

Some other relations that are useful regarding the translation operators are

$$e^{i\tau D}f(x)\varphi(x) = [e^{i\tau D}f(x)][e^{i\tau D}\varphi(x)]. \tag{2.37}$$

Also,

$$[e^{i\tau D}, x] = \tau e^{i\tau D} \tag{2.38}$$

and more generally

$$[e^{i\tau D}, g(x)] = [g(x+\tau) - g(x)]e^{i\tau D}. \tag{2.39}$$

The adjoint of the translation operator is given by

$$(e^{i\tau D})^\dagger = e^{-i\tau D}. \tag{2.40}$$

This can be proved in the following way. We have

$$\int g^* e^{i\tau D} f \, dx = \int g^*(x) f(x+\tau) \, dx = \int g^*(x-\tau) f(x) \, dx$$
$$= \int f(x) \left[e^{-i\tau D} g(x) \right]^* dx \qquad (2.41)$$

which shows Eq. (2.40).

2.4 The Operator $e^{i\theta X + i\tau D}$

The simplification and disentanglement of the operator e^{A+B} is very important and also very difficult. We will be discussing this in detail in a later chapter; however, fortunately, there are some special cases where simplification is possible. When the two operators, A and B, both commute with their commutator

$$[A, [A, B]] = [B, [A, B]] = 0, \qquad (2.42)$$

then

$$e^{A+B} = e^{-\frac{1}{2}[A,B]} e^A e^B = e^{\frac{1}{2}[A,B]} e^B e^A. \qquad (2.43)$$

This is a special case of the Baker-Cambell-Hausdorf formula. See Eq. (15.50).

Of particular interest is the operator $e^{i\theta X + i\tau D}$ with real θ and τ. The relevant commutator is

$$[i\theta X, i\tau D] = -i\theta\tau \qquad (2.44)$$

and using Eq. (2.43) we have

$$e^{i\theta X + i\tau D} = e^{i\theta\tau/2} e^{i\theta X} e^{i\tau D} = e^{-i\theta\tau/2} e^{i\tau D} e^{i\theta X}. \qquad (2.45)$$

Therefore for an arbitrary function, $f(x)$, we have

$$e^{i\theta X + i\tau D} f(x) = e^{i\theta\tau/2} e^{i\theta X} e^{i\tau D} f(x) = e^{i\theta\tau/2} e^{i\theta x} f(x+\tau). \qquad (2.46)$$

In addition,

$$e^{i\theta X} e^{i\tau D} = e^{-i\theta\tau} e^{i\tau D} e^{i\theta X} \qquad (2.47)$$

and

$$e^{i\tau D} e^{i\theta X} = e^{i\theta\tau} e^{i\theta X} e^{i\tau D}. \qquad (2.48)$$

Also,

$$e^{i\theta X + i\tau D} f(x) g(x) = [e^{i\theta X + i\tau D} f(x)][e^{i\tau D} g(x)]. \qquad (2.49)$$

Characteristic function operator. It is convenient to define the characteristic function operator for the Weyl correspondence

$$\mathcal{M}(\theta, \tau) = e^{i\theta X + i\tau D}. \qquad (2.50)$$

2.4. The Operator $e^{i\theta X+i\tau D}$

The reason why this is called the characteristic function operator will become clear in Chap. 5. It often arises that one has to evaluate the product of two such operators $\mathcal{M}(\theta',\tau')\mathcal{M}(\theta,\tau)$. Explicitly

$$\mathcal{M}(\theta',\tau')\mathcal{M}(\theta,\tau) = e^{i\theta'X+i\tau'D}e^{i\theta X+i\tau D} \tag{2.51}$$

$$= e^{i\theta'\tau'/2}\, e^{i\theta\tau/2}\, e^{i\theta\tau'}\, e^{i(\theta+\theta')X}\, e^{i(\tau'+\tau)D}. \tag{2.52}$$

But using Eq. (2.45) we have that

$$e^{i(\theta+\theta')X}e^{i(\tau'+\tau)D} = e^{-i(\theta+\theta')(\tau+\tau')/2}e^{i(\theta+\theta')X+i(\tau+\tau')D} \tag{2.53}$$

and therefore we obtain

$$e^{i\theta'\tau'/2}\, e^{i\theta\tau/2}\, e^{i\theta\tau'}\, e^{i(\theta+\theta')X}\, e^{i(\tau'+\tau)D}$$
$$= e^{i\theta'\tau'/2}\, e^{i\theta\tau/2}\, e^{i\theta\tau'}\, e^{-i(\theta+\theta')(\tau+\tau')/2}e^{i(\theta+\theta')X+i(\tau+\tau')D} \tag{2.54}$$

giving

$$e^{i\theta'X+i\tau'D}e^{i\theta X+i\tau D} = e^{i(\theta\tau'-\theta'\tau)/2}e^{i(\theta+\theta')X+i(\tau+\tau')D} \tag{2.55}$$

or

$$\mathcal{M}(\theta',\tau')\mathcal{M}(\theta,\tau) = e^{i(\theta\tau'-\theta'\tau)/2}\,\mathcal{M}(\theta+\theta',\tau+\tau'). \tag{2.56}$$

For future reference we also write

$$e^{i\theta X+i\tau D}e^{i\theta'X+i\tau'D} = e^{i(\theta'\tau-\theta\tau')/2}e^{i(\theta+\theta')X+i(\tau+\tau')D}. \tag{2.57}$$

One also has

$$\mathcal{M}(\theta-\theta',\tau-\tau')\mathcal{M}(\theta',\tau') = e^{i(\theta\tau'-\theta'\tau)/2}\,\mathcal{M}(\theta,\tau). \tag{2.58}$$

Also of interest is the commutator

$$[\theta'X+\tau'D, \theta X+\tau D] = \theta'\tau[X,D]+\theta\tau'[D,X] = i(\theta'\tau-\theta\tau'). \tag{2.59}$$

The adjoint of $e^{i\theta X+i\tau D}$. The adjoint of the characteristic function operator often arises. We designate it by $\mathcal{M}^\dagger(\theta,\tau)$ and we now show that it is given by

$$\mathcal{M}^\dagger(\theta,\tau) = \mathcal{M}(-\theta,-\tau) = e^{-i\theta X-i\tau D}. \tag{2.60}$$

To prove this consider

$$\int g^* e^{i\theta X+i\tau D} f\, dx = \int g^*(x) e^{i\theta\tau/2} e^{i\theta x} e^{i\tau D} f(x)\, dx = \int g^*(x) e^{i\theta\tau/2} e^{i\theta x} f(x+\tau)\, dx. \tag{2.61}$$

A straightforward change of variables results in

$$\int g^* e^{i\theta X + i\tau D} f \, dx = \int f(x) \left[e^{i\theta\tau/2} e^{-i\theta x} g(x-\tau) \right]^* dx \qquad (2.62)$$

$$= \int f(x) \left[e^{i\theta\tau/2} e^{-i\theta x} e^{-i\tau D} g(x-\tau) \right]^* dx \qquad (2.63)$$

$$= \int f(x) \left[e^{-i\theta X - i\tau D} g(x) \right]^* dx \qquad (2.64)$$

which proves Eq. (2.60).

Generalized characteristic function operator. In later chapters, and particulalry in Chap. 4, we will study the generalized characteristic function operator given by

$$\mathcal{M}^\Phi(\theta, \tau) = \Phi(\theta, \tau) \, e^{i\theta X + i\tau D} \qquad (2.65)$$

where $\Phi(\theta, \tau)$ is a complex function called the kernel. The adjoint of this operator is

$$\mathcal{M}^{\Phi\dagger}(\theta, \tau) = \Phi^*(\theta, \tau) \, e^{-i\theta X - i\tau D}. \qquad (2.66)$$

2.5 The Operator $e^{\xi H} A e^{-\xi H}$

The operator $e^{\xi H} A e^{-\xi H}$ comes up often in many fields and is particularly important in quantum mechanics. An expansion for this operator is

$$e^{\xi H} A e^{-\xi H} = A + \xi[H, A] + \frac{1}{2!}\xi^2[H,[H,A]] + \frac{1}{3!}\xi^3[H,[H,[H,A]]] + \cdots . \qquad (2.67)$$

The reason the operator is fundamental in quantum mechanics is because it is the formal solution to the Heisenberg equation of motion for an operator. Heisenberg's equation of motion for a time independent Hamiltonian operator, H, and for an operator A that does not explicitly depend on time is,

$$\frac{dA}{dt} = i[H, A]. \qquad (2.68)$$

The formal solution of this equation is

$$A(t) = e^{itH} A(0) e^{-itH} \qquad (2.69)$$

as can be readily verified.

An example of Eq. (2.67) is the simplification of $e^{i\beta D} X e^{-i\beta D}$. Since $[D, X] = -i$ the series truncates after the second term and we have that

$$e^{i\beta D} X e^{-i\beta D} = X - i(i\beta) = X + \beta. \qquad (2.70)$$

2.6. Phase-Space Operator Formulas

Repeated Commutator. Expressions such as $[H, [H, [H, A]]]$ are called repeated commutators and appear often. A convenient notation to denote them is $[H, A]_n$ where n indicates the number of repetitions of the commutator. Eq. (2.67) may now be written as

$$e^{\xi H} A e^{-\xi H} = A + \sum_{n=1}^{\infty} \frac{1}{n!} \xi^n [H, A]_n. \quad (2.71)$$

An explicit expression for the repeated commutator is [86]

$$[H, A]_n = \sum_{k=0}^{n} \binom{n}{k} (-1)^{n-k} H^k A H^{n-k}. \quad (2.72)$$

Also, we note that

$$[H, A]_{n+1} = H[H, A]_n - [H, A]_n H. \quad (2.73)$$

2.6 Phase-Space Operator Formulas

In dealing with operators in phase-space one comes across integrals that can be written in an operational form that allows one to manipulate operators in very advantageous ways. We list some of the important ones:

i) If $f(\theta, \tau)$ and $\hat{a}(\theta, \tau)$ are arbitrary functions then

$$\iint f(\theta, \tau) \hat{a}(\theta, \tau) e^{i\theta x + i\tau p} \, d\theta \, d\tau = f\left(\frac{1}{i}\frac{\partial}{\partial x}, \frac{1}{i}\frac{\partial}{\partial p}\right) a(x, p) \quad (2.74)$$

where the normalization in the Fourier transform, $\hat{a}(\theta, \tau)$, is as in Eq. (1.13).
ii) For any two functions $f(\theta, \tau)$ and $F(x, p)$,

$$\frac{1}{4\pi^2} \iint f(\theta, \tau) e^{i\theta(x'-x)+i\tau(p'-p)} F(x', p') \, d\theta \, d\tau \, dx' \, dp' = f\left(i\frac{\partial}{\partial x}, i\frac{\partial}{\partial p}\right) F(x, p). \quad (2.75)$$

iii) For any two functions $g(x)$ and $h(x, p)$,

$$g\left(x + \tfrac{i}{2}\frac{\partial}{\partial p}\right) h(x, p) = \frac{1}{2\pi} \iint g(x') e^{i\theta(x-x')} h(x, p - \tfrac{1}{2}\theta) \, d\theta \, dx' \quad (2.76)$$

$$= \frac{1}{\pi} \iint g(x') e^{2i(p-p')(x-x')} h(x, p') \, dp' \, dx'. \quad (2.77)$$

iv) If we have two phase-space functions $a(x, p)$ and $b(x, p)$ then

$$f\left(i\frac{\partial}{\partial x}, i\frac{\partial}{\partial p}\right) a(x, p) b(x, p) = f\left(i\frac{\partial}{\partial x_a} + i\frac{\partial}{\partial x_b}, i\frac{\partial}{\partial p_a} + i\frac{\partial}{\partial p_b}\right) a(x, p) b(x, p). \quad (2.78)$$

In Eq. (2.78) the meaning of $\frac{\partial}{\partial x_a}$ is that it operates only on $a(x,p)$ and similarly for the other partial derivatives.

v) If $f(x,p)$ is a real function then

$$\int a(x,p) f\left(i\frac{\partial}{\partial x}, i\frac{\partial}{\partial p}\right) b(x,p)\, dx\, dp = \int b(x,p) f\left(\frac{1}{i}\frac{\partial}{\partial x}, \frac{1}{i}\frac{\partial}{\partial p}\right) a(x,p)\, dx\, dp. \tag{2.79}$$

vi) For any function of two variables, say $a(x,p)$,

$$\frac{1}{\pi}\iint a(x'+x, p'+p)\, e^{i2x'p'}\, dx'\, dp' = \exp\left[-\frac{1}{2i}\frac{\partial}{\partial x}\frac{\partial}{\partial p}\right] a(x,p). \tag{2.80}$$

vii) For any two functions $a(x,p)$ and $f(x,p)$

$$a\left(x + \frac{i}{2}\frac{\partial}{\partial p}, p - \frac{i}{2}\frac{\partial}{\partial x}\right) f(x,p) = \iint \widehat{a}(\theta, \tau)\, f(x+\tau/2, p-\theta/2) e^{i\theta x + i\tau p}\, d\theta\, d\tau. \tag{2.81}$$

2.7 Transforms and the Representation of Functions

The writing of a function in different domains is a fundamental idea that has been developed over the last few hundred years but is particularly crucial in many physical theories and quantum mechanics in particular. Perhaps the first idea along these lines is the Taylor series but certainly the most important historically, is the expression of a function in the Fourier domain. However, there are an infinite number of domains that one can express a function in and in fact any Hermitian operator generates a domain, as we will now discuss. The remarkable historical event that necessitated the development of the expansion of functions in a different domain was the discovery of quantum mechanics. We will describe why this is so later but we first discuss some of the standard reasons. First is that the function may be more simply characterized in another representation and hence gives considerably more insight into the nature of the function. Thus, for example, a messy function in one domain may have a simple expression in the Fourier domain or the Hermite function representation, etc. Secondly, if we want to construct functions with certain characteristics which are described in a certain domain, then clearly we should do the analysis in the representation of that domain and then transform back to the working domain. For example if we want a function that has spatial frequencies only in a band, then clearly the function should be constructed in the frequency domain and then transformed back to the spatial domain.

Representation of functions in domains. Any Hermitian operator generates a domain or representation as we now discuss. One solves the eigenvalue problem for the operator and that results in either continuous or discrete eigenvalues. We write

2.7. Transforms and the Representation of Functions

these two cases as

$$Au(\lambda, x) = \lambda u(\lambda, x) \quad \text{continuous spectrum,} \qquad (2.82)$$
$$Au_n(x) = \lambda_n u_n(x) \quad \text{discrete spectrum,} \qquad (2.83)$$

where λ or λ_n are the continuous or discrete eigenvalues respectively and where the u's are the corresponding eigenfunctions. In writing Eqs. (2.82) and (2.83) we have assumed that the operator is expressed in the x representation; however the eigenvalue problem can be solved in any representation. The eigenfunctions thus generated are complete and orthogonal and that allows one to express any function in the representation defined by the operator. We deal with the continuous case and discrete case separately.

Continuous case. By complete and orthogonal one means that the eigenfunctions satisfy

$$\int u^*(\lambda', x)\, u(\lambda, x)\, dx = \delta(\lambda - \lambda'), \qquad (2.84)$$

$$\int u^*(\lambda, x')\, u(\lambda, x)\, d\lambda = \delta(x - x'). \qquad (2.85)$$

Eq. (2.84) is called delta function normalization and Eq. (2.85) is called the closure relation. Any function can be expanded as

$$\varphi(x) = \int F(\lambda) u(\lambda, x)\, d\lambda \qquad (2.86)$$

where the inverse transformation is given by

$$F(\lambda) = \int \varphi(x) u^*(\lambda, x)\, dx. \qquad (2.87)$$

The function $F(\lambda)$ is called the *transform* of $f(x)$ and may be considered as the representation of the function in the A representation.

Discrete Case. If the spectrum is discrete the eigenfunctions are orthogonal and it is standard to normalize them so that

$$\int u_m^*(x) u_n(x)\, dx = \delta_{nm} \qquad (2.88)$$

Also, one has that

$$\sum_n u_n^*(x') u_n(x) = \delta(x' - x) \qquad (2.89)$$

where the summation runs over all the eigenfunctions. Any function can be expanded as

$$\varphi(x) = \sum_n c_n u_n(x) \qquad (2.90)$$

where the coefficients c_n are given by

$$c_n = \int u_n^*(x)\varphi(x)\,dx. \tag{2.91}$$

The set of coefficients $\{c_n\}$ can be thought of as the function in the u_n representation or as the discrete transform of the function.

Functions of operators operating on an eigenfunction. A particulary important result is that if we have a function of an operator, $f(A)$, then

$$f(A)u(\lambda, x) = f(\lambda)u(\lambda, x) \tag{2.92}$$

where $u(\lambda, x)$ are the eigenfunctions of A. This can be proven as follows. Using Eq. (2.22) for the definition of a function of an operator

$$f(A) = \sum_{n=0}^{\infty} f_n A^n \tag{2.93}$$

we have

$$f(A)u(\lambda, x) = \sum_{n=0}^{\infty} f_n A^n u(\lambda, x) = \sum_{n=0}^{\infty} f_n \lambda^n u(\lambda, x) = f(\lambda)u(\lambda, x). \tag{2.94}$$

This allows one to evaluate the operation of $f(A)$ on an arbitrary function, $\varphi(x)$. Using Eq. (2.86) we write

$$\varphi(x) = \int F(\lambda)u(\lambda, x)\,d\lambda \tag{2.95}$$

where the transform, $F(\lambda)$, is given by

$$F(\lambda) = \int \varphi(x)u^*(\lambda, x)\,dx. \tag{2.96}$$

Therefore,

$$f(A)\varphi(x) = \int F(\lambda)f(A)u(\lambda, x)\,d\lambda = \int F(\lambda)f(\lambda)u(\lambda, x)\,d\lambda. \tag{2.97}$$

If we further substitute for $F(\lambda)$ then

$$f(A)\varphi(x) = \iint \varphi(x')u^*(\lambda, x')f(\lambda)u(\lambda, x)\,d\lambda dx' \tag{2.98}$$

which can be written as

$$f(A)\varphi(x) = \int \varphi(x')r(x', x)dx' \tag{2.99}$$

2.8. Delta Function

where
$$r(x', x) = \int u^*(\lambda, x') f(\lambda) u(\lambda, x)\, d\lambda. \tag{2.100}$$

If the eigenfunctions are discrete as in Eq. (2.83) then
$$f(A) u_n(x) = f(\lambda_n) u_n(x) \tag{2.101}$$

and
$$f(A)\varphi(x) = \int \sum_{n=0}^{\infty} u_n^*(x') f(c_n) u_n(x)\, \varphi(x')\, dx' \tag{2.102}$$
$$= \int \varphi(x') r(x', x)\, dx' \tag{2.103}$$

where now
$$r(x', x) = \sum_{n=0}^{\infty} u_n^*(x') f(\lambda_n) u_n(x). \tag{2.104}$$

Example. For the D operator the eigenfunctions are
$$u(\lambda, x) = \frac{1}{\sqrt{2\pi}} e^{i\lambda x} \tag{2.105}$$

and hence for any function $f(D)$ we have
$$f(D) e^{i\lambda x} = f(\lambda) e^{i\lambda x}. \tag{2.106}$$

Furthermore, suppose we have a function, $\varphi(x)$, then, using Eq. (2.99) and Eq. (2.100)
$$f(D)\varphi(x) = \frac{1}{2\pi} \iint \varphi(x') f(\lambda) e^{-i\lambda(x'-x)}\, d\lambda\, dx' \tag{2.107}$$
$$= \frac{1}{\sqrt{2\pi}} \int \varphi(x') \widehat{f}(x' - x)\, dx' \tag{2.108}$$

where, as usual, the Fourier transform is given by
$$\widehat{f}(x) = \frac{1}{\sqrt{2\pi}} \int f(\lambda) e^{-i\lambda x}\, d\lambda. \tag{2.109}$$

2.8 Delta Function

We will be using the delta function freely and it is worthwhile to list some of its basic properties. The fundamental representation of the delta function is
$$\delta(x - \xi) = \frac{1}{2\pi} \int_{-\infty}^{\infty} e^{\pm iy(x-\xi)}\, dy \tag{2.110}$$

and its basic property is that, for a function $f(x)$,

$$\int_{-\infty}^{\infty} f(x)\delta(x-\xi)\,dx = f(\xi). \tag{2.111}$$

If the integration is one sided in Eq. (2.110) then

$$\int_{0}^{\infty} e^{iy(x-\xi)}\,dy = \pi\,\delta(x-\xi) + \frac{i}{x-\xi} \tag{2.112}$$

where the integration implies taking the principle part. The delta function has the following symbolic relations:

$$\delta(x) = \delta(-x), \tag{2.113}$$
$$x\delta(x) = 0, \tag{2.114}$$
$$f(x)\delta(x-\xi) = f(\xi)\delta(x-\xi), \tag{2.115}$$
$$\delta(\xi x) = \xi^{-1}\delta(x), \qquad \xi > 0, \tag{2.116}$$
$$\int_{-\infty}^{\infty} f(x)\frac{d^n}{dx^n}\delta(x-\xi)\,dx = (-1)^n \frac{d^n}{d\xi^n} f(\xi). \tag{2.117}$$

A particularly important relation is that

$$\delta(g(x)) = \sum_{i} \frac{1}{|g'(x_i)|} \delta(x - x_i) \tag{2.118}$$

where x_i are the zeros of $g(x)$ and $g'(x_i)$ is the derivative $g(x)$ evaluated at the zero's. By a symbolic relation we mean that if both sides of the relation are multiplied by an arbitrary function and integrated from $-\infty$ to ∞, an identity is obtained. For example, when we say that $\delta(x) = \delta(-x)$ what it means is that for an arbitrary function $f(x)$,

$$\int_{-\infty}^{\infty} f(x)\delta(x)\,dx = \int_{-\infty}^{\infty} f(x)\delta(-x)\,dx. \tag{2.119}$$

Two other important representations of the delta function are

$$\delta(x) = \frac{1}{2}\frac{d^2}{dx^2}|x| \tag{2.120}$$

and

$$\delta(x) = \frac{d}{dx}\eta(x) \tag{2.121}$$

where $\eta(x)$ is the step function

$$\eta(x) = \begin{cases} 1 & x \geq 0 \\ 0 & x < 0. \end{cases} \tag{2.122}$$

We point out that
$$x\delta(x-\xi) = \xi\delta(x-\xi) \qquad (2.123)$$
and hence $\delta(x-\xi)$ are the eigenfunctions of x, with eigenvalues ξ, which range from $-\infty$ to ∞. In fact it was to solve the eigenvalue problem for x as exemplified by Eq. (2.123) that Dirac invented the delta function. In quantum mechanics this is paramount because x is the position operator and the eigenvalues are the measurable quantities. What Eq. (2.123) shows is that the measurable values for position, the eigenvalues ξ, are continuous since ξ may be any real number. Therefore one says that position is not quantized.

2.9 Intensity, Probability, and Averages

The transform $F(\lambda)$ as given by Eq. (2.87) gives an indication of how important a particular value of λ is for the function $\varphi(x)$. In particular, one takes the density or intensity of λ to be $|F(\lambda)|^2$. In the discrete case one takes $|c_n|^2$ to be the intensity for λ_n. In quantum mechanics $|F(\lambda)|^2$ is the probability distribution for λ and if the spectrum is discrete then $|c_n|^2$ is the probability for obtaining λ_n.

For this interpretation to be consistent one has to show that intensity is preserved. In particular for the continuous case we have

$$\int |\varphi(x)|^2 dx = \int |F(\lambda)|^2 d\lambda. \qquad (2.124)$$

This is easily proven. Now consider the average defined by

$$\langle \lambda \rangle = \int \lambda |F(\lambda)|^2 d\lambda. \qquad (2.125)$$

It is a remarkable fact that $\langle \lambda \rangle$ can be calculated in the x domain directly by way of

$$\langle \lambda \rangle = \int \varphi^*(x) A \varphi(x)\, dx \qquad (2.126)$$

and furthermore for a real function, $g(\lambda)$, its average is

$$\int g(\lambda)|F(\lambda)|^2 d\lambda = \int \varphi^*(x) g(A) \varphi(x)\, dx. \qquad (2.127)$$

For the discrete case we have

$$\int |\varphi(x)|^2 dx = \sum_n |c_n|^2 \qquad (2.128)$$

and

$$\langle \lambda \rangle = \sum_n \lambda_n |c_n|^2 = \int \varphi^*(x) A \varphi(x)\, dx. \qquad (2.129)$$

Furthermore $\langle g(\lambda) \rangle$ is given by

$$\langle g(\lambda) \rangle = \sum_n g(\lambda_n) |c_n|^2 = \int \varphi^*(x) g(A) \varphi(x) \, dx. \qquad (2.130)$$

The proof of these important statements will be given in Chap. 12.

Probability interpretation. Suppose we assume that $|\varphi(x)|^2$ is the probability for x and that it is appropriately normalized so that

$$\int |\varphi(x)|^2 \, dx = 1. \qquad (2.131)$$

Notice now that we also have

$$\int |F(\lambda)|^2 \, d\lambda = 1 \qquad (2.132)$$

and therefore $|F(\lambda)|^2$ can be interpreted as the probability of measuring λ. This is the probabilistic interpretation of quantum mechanics where $\varphi(x)$ is called the wave function or state function and the λ's are the numerical values for the physical quantity represented by the operator A. For the discrete case we have that

$$\sum_n |c_n|^2 = 1 \qquad (2.133)$$

and then we say that $|c_n|^2$ is the probability of obtaining λ_n. Since λ_n are discrete one says that the numerical values possible for the physical quantity represented by the operator A are quantized.

Notice that the left hand side of Eq. (2.127) and Eq. (2.130) are the standard definitions of averages in standard probability theory for the continuous and discrete case respectively. That these quantities can be calculated by the right hand side of the respective equations is something that is proved and we will do so in Chap. 12. What is remarkable is that these methods of calculating averages arose naturally in quantum mechanics.

2.10 Simplifying Functions of Operators

An important issue is the consideration of a function of operators, say $H(X, D)$, where one requires the simplification of $G(H(X, D))$ where G is another function. For example, suppose we want to simplify $(X + D)^n$. In our notation $H(X, D) = X + D$ and $G(H) = H^n$. There are many ways to do this and brute force, using the commutation relation, often works. We give here one method [14, 95] that depends on solving the eigenvalue problem

$$H(X, D) \, u(\lambda, x) = \lambda \, u(\lambda, x). \qquad (2.134)$$

2.10. Simplifying Functions of Operators

The solution to this eigenvalue problem gives rise to a complete set of eigenfunctions, as given by Eq. $u(\lambda, x)$ and hence for any function $f(x)$,

$$f(x) = \int u(\lambda, x) F(\lambda) \, d\lambda \tag{2.135}$$

with

$$F(\lambda) = \int u^*(\lambda, x) f(x) \, dx. \tag{2.136}$$

Now consider

$$G(H(X, D)) f(x) = G(H(X, D)) \int u(\lambda, x) F(\lambda) \, d\lambda \tag{2.137}$$

$$= \int G(H(X, D)) u(\lambda, x) F(\lambda) \, d\lambda \tag{2.138}$$

$$= \int G(\lambda) u(\lambda, x) F(\lambda) \, d\lambda. \tag{2.139}$$

Substituting for $F(\lambda)$ we have

$$G(H(X, D)) f(x) = \iint G(\lambda) u^*(\lambda, x') u(\lambda, x) f(x') \, dx' \, d\lambda \tag{2.140}$$

which can be written as

$$G(H(X, D)) f(x) = \int K(x', x) f(x') \, dx' \tag{2.141}$$

with

$$K(x', x) = \int G(\lambda) u^*(\lambda, x') u(\lambda, x) \, d\lambda. \tag{2.142}$$

If the spectrum is discrete then the same steps lead to Eq. (2.141) where now

$$K(x', x) = \sum_n G(\lambda_n) u_n^*(x') u_n(x). \tag{2.143}$$

Example. Suppose we seek the simplification of $e^{i\theta X + i\tau D}$ as was discussed in Sec. 2.4. As per Eq. (2.134) we have to solve the eigenvalue problem

$$(\theta X + \tau D) u(\lambda, x) = \lambda u(\lambda, x) \tag{2.144}$$

or explicitly in the x representation,

$$\left(\theta x - i\tau \frac{d}{dx} \right) u(\lambda, x) = \lambda u(\lambda, x). \tag{2.145}$$

The solution normalized to a delta function is

$$u(\lambda, x) = \frac{1}{\sqrt{2\pi\tau}} e^{i(\lambda x - \theta x^2/2)/\tau}. \tag{2.146}$$

For a function, $\varphi(x)$, the transform, $F(\lambda)$, is then

$$F(\lambda) = \frac{1}{\sqrt{2\pi\tau}} \int \varphi(x) e^{-i(\lambda x - \theta x^2/2)/\tau} dx \tag{2.147}$$

and the inverse transformation is

$$\varphi(x) = \frac{1}{\sqrt{2\pi\tau}} \int F(\lambda) e^{i(\lambda x - \theta x^2/2)/\tau} d\lambda. \tag{2.148}$$

According to Eq. (2.142) we have

$$e^{i\theta X + i\tau D} \varphi(x) = \int K(x', x) \varphi(x') \, dx' \, d\lambda \tag{2.149}$$

with

$$K(x', x) = \int e^{i\lambda} u^*(\lambda, x') u(\lambda, x) \, d\lambda \, dx' \tag{2.150}$$

$$= \frac{1}{2\pi\tau} \int e^{i\lambda} e^{-i(\lambda x' - \theta x'^2/2)/\tau} e^{i(\lambda x - \theta x^2/2)/\tau} \, d\lambda \tag{2.151}$$

$$= \delta(x' - \tau - x) e^{i\theta(x'^2 - x^2)/2\tau}. \tag{2.152}$$

Therefore

$$e^{i\theta X + i\tau D} \varphi(x) = \int \delta(x' - \tau - x) e^{i\theta(x'^2 - x^2)/2\tau} \varphi(x') \, dx' \tag{2.153}$$

$$= e^{i\theta(\tau^2 + 2\tau x)/2\tau} \varphi(\tau + x) \tag{2.154}$$

$$= e^{i\theta\tau/2} e^{i\theta x} \varphi(\tau + x) \tag{2.155}$$

$$= e^{i\theta\tau/2} e^{i\theta x} e^{i\tau D} \varphi(x). \tag{2.156}$$

Hence, we can write

$$e^{i\theta X + i\tau D} = e^{i\theta\tau/2} e^{i\theta X} e^{i\tau D} \tag{2.157}$$

which is Eq. (2.48)

2.11 Rearrangement of Operators

If we have a function of operators $G(X, D)$ we may want to rearrange it so that all the D factors are to the right of the X factors or the other way around. A

2.11. Rearrangement of Operators

procedure that can be used is the following [14, 95]. Write Eq. (2.140) as

$$G(X,D)\varphi(x) = \iint G(\lambda)\, u^*(\lambda, x')\, u(\lambda, x)\varphi(x')\, dx'\, d\lambda \qquad (2.158)$$

$$= \iint G(\lambda)\, u^*(\lambda, x' + x)\, u(\lambda, x)\varphi(x' + x)\, dx'\, d\lambda \qquad (2.159)$$

$$= \iint G(\lambda)\, u^*(\lambda, x' + x)\, u(\lambda, x)\, e^{ix'D}\varphi(x)\, dx'\, d\lambda \qquad (2.160)$$

and therefore we have

$$G(X,D) = \iint G(\lambda)\, u^*(\lambda, x' + X)\, u(\lambda, X)\, e^{ix'D}\, dx'\, d\lambda. \qquad (2.161)$$

If the spectrum is discrete then

$$G(X,D) = \sum_n G(\lambda_n)\, u_n^*(x' + X)\, u_n(X)\, e^{ix'D}\, dx'. \qquad (2.162)$$

Notice that in Eq. (2.161) all the X factors are to the left of the D factors. Therefore one way to evaluate the integral in Eq. (2.161) is to replace X and D by ordinary variables, say x and p,

$$G(x,p) = \sum_n G(\lambda_n)\, u_n^*(x' + x)\, u_n(x)\, e^{ix'p}\, dx' \qquad (2.163)$$

do the integral and then arrange the expression so that all the x factors are to the left of the p factors; then substitute X and D for x and p respectively. This procedure will be illustrated in subsequent chapters. Here we give one example. Suppose we want to expand

$$G(X,D) = (X+D)^n \qquad (2.164)$$

and rearrange it so that all the X factors are to the left of the D factors. That can be done by brute force using the commutator relations but one would soon be entangled in laborious algebra. We now show how that can be done using the above method. The eigenvalue problem is

$$\left(x - i\frac{d}{dx}\right) u(\lambda, x) = \lambda\, u(\lambda, x) \qquad (2.165)$$

and the solutions are

$$u(\lambda, x) = \frac{1}{\sqrt{2\pi}}\, e^{i(\lambda x - x^2/2)}. \qquad (2.166)$$

Therefore using Eq. (2.161) we have

$$G(x,p) = \frac{1}{2\pi}\iint \lambda^n\, e^{-i(\lambda(x'+x)-(x'+x)^2)/2)}\, e^{i(\lambda x - x^2/2)}\, e^{ix'p}\, dx'\, d\lambda \qquad (2.167)$$

$$= \frac{1}{2\pi}\iint \lambda^n\, e^{-ix'(\lambda - x - p) + ix'^2/2}\, dx'\, d\lambda \qquad (2.168)$$

which simplifies to

$$G(x,p) = \frac{(\sqrt{2i})^n}{\sqrt{\pi}} \int (iy + \frac{1}{\sqrt{2i}}(x+p))^n e^{-y^2} dy. \tag{2.169}$$

Now, the Hermite polynomials, $H_n(x)$, are

$$H_n(x) = \sum_{k=0}^{[n/2]} \frac{(-1)^k n!}{k!(n-2k)!} x^{n-2k} \tag{2.170}$$

where $[n/2]$ is the greatest integer function. Pertinent to our considerations is that

$$H_n(x) = \frac{2^n}{\sqrt{\pi}} \int (x+iy)^n e^{-y^2} dy. \tag{2.171}$$

Therefore we have

$$G(x,p) = \left(\frac{i}{2}\right)^{n/2} H_n(\frac{1}{\sqrt{2i}}(x+p)) \tag{2.172}$$

and

$$G(x,p) = \left(\frac{i}{2}\right)^{n/2} \sum_{k=0}^{[n/2]} \frac{(-1)^k n! \left(2\frac{x+p}{\sqrt{2i}}\right)^{n-2k}}{k!(n-2k)!}. \tag{2.173}$$

Expanding $(x+p)^{n-2k}$ in a binomial series one finally obtains that

$$G(x,p) = \sum_{k=0}^{[n/2]} \sum_{l=0}^{n-2k} \left(\frac{1}{2i}\right)^k \frac{n!}{l!k!(n-2k-l)!} x^{n-2k-l} p^l. \tag{2.174}$$

Now all the x factors are to the left of the p factors and we can write

$$G(X,D) = (X+D)^n = \sum_{k=0}^{[n/2]} \sum_{l=0}^{n-2k} \left(\frac{1}{2i}\right)^k \frac{n!}{l!k!(n-2k-l)!} X^{n-2k-l} D^l. \tag{2.175}$$

Chapter 3

The Weyl Operator

3.1 The Weyl Correspondence

There are a number of ways one can define the Weyl operator but we choose the standard one and indeed it is the way Weyl did it [89]. The fundamental idea is to associate an operator, $A(X, D)$, with the ordinary function, $a(x, p)$, called a symbol or c-function. We express the Weyl association symbolically by

$$A(X, D) \leftrightarrow a(x, p). \tag{3.1}$$

As discussed in Chap. 1 the two variables, x and p are ordinary variables while X and D are operators where

$$X = \begin{cases} x & \text{in the } x \text{ representation} \\ i\frac{d}{dp} & \text{in the Fourier representation} \end{cases} \tag{3.2}$$

and

$$D = \begin{cases} \frac{1}{i}\frac{d}{dx} & \text{in the } x \text{ representation} \\ p & \text{in the Fourier representation.} \end{cases} \tag{3.3}$$

They satisfy the basic commutation relation

$$[X, D] = i. \tag{3.4}$$

The Weyl rule is the following. For a symbol, $a(x, p)$, define its Fourier transform by

$$\widehat{a}(\theta, \tau) = \frac{1}{4\pi^2} \iint a(x, p) \, e^{-i\theta x - i\tau p} \, dx \, dp \tag{3.5}$$

in which case

$$a(x, p) = \iint \widehat{a}(\theta, \tau) \, e^{i\theta x + i\tau p} \, d\theta \, d\tau. \tag{3.6}$$

The Weyl operator, $A(X, D)$, *corresponding* to, or *associated* with, $a(x, p)$ is defined by the substitution of the operators X and D for x and p in Eq. (3.6),

$$A(X, D) = \iint \hat{a}(\theta, \tau) e^{i\theta X + i\tau D} d\theta \, d\tau. \tag{3.7}$$

We call $A(X, D)$ the Weyl operator. Simplification of $A(X, D)$ is possible in a number of ways. One is the disentanglement of $e^{i\theta X + i\tau D}$ as we discuss now and the other is using the rearrangement procedure that will be discussed in Sec. 3.4. Since

$$e^{i\theta X + i\tau D} = e^{i\theta \tau / 2} e^{i\theta X} e^{i\tau D} = e^{-i\theta \tau / 2} e^{i\tau D} e^{i\theta X} \tag{3.8}$$

we have that

$$A(X, D) = \iint \hat{a}(\theta, \tau) e^{i\theta \tau / 2} e^{i\theta X} e^{i\tau D} d\theta \, d\tau \tag{3.9}$$

and

$$A(X, D) = \iint \hat{a}(\theta, \tau) e^{-i\theta \tau / 2} e^{i\tau D} e^{i\theta X} d\theta \, d\tau. \tag{3.10}$$

Equivalently, substitute Eq. (3.5) into Eq. (3.9) or into (3.10) to obtain

$$A(X, D) = \frac{1}{4\pi^2} \iiiint a(x, p) e^{i\theta(X - x) + i\tau(D - p)} d\theta \, d\tau \, dx \, dp \tag{3.11}$$

$$= \frac{1}{4\pi^2} \iiiint a(x, p) e^{i\theta \tau / 2} e^{i\theta(X - x)} e^{i\tau(D - p)} d\theta \, d\tau \, dx \, dp. \tag{3.12}$$

It is important to appreciate that in the above equations both X and D are operators. Whether we use the pair (x, D) or (X, p) depends on whether we operate on functions of x or functions in the Fourier domain. The examples below will illustrate these different situations. Furthermore, $A(X, D)$ can operate on functions expressed in any representation.

3.2 Operation on a Function

We now consider the operation of the Weyl operator on an arbitrary function, $\varphi(x)$, and use $A[\varphi]$ to signify it,

$$A[\varphi] = A(X, D) \, \varphi(x). \tag{3.13}$$

Of course $A[\varphi]$ is a function of x and often it is written as $A[\varphi](x)$ but we will simply write $A[\varphi]$ for the sake of notational clarity, as the dependence on the variable will be clear from the context. We have

$$A(X, D) \, \varphi(x) = \iint \hat{a}(\theta, \tau) e^{i\theta \tau / 2} e^{i\theta x} e^{i\tau D} \varphi(x) \, d\theta \, d\tau \tag{3.14}$$

$$= \iint \hat{a}(\theta, \tau) e^{i\theta \tau / 2} e^{i\theta x} \varphi(x + \tau) \, d\theta \, d\tau \tag{3.15}$$

3.4. Operational Form

where in going from Eq. (3.14) to Eq. (3.15) we have used the fact that $e^{i\tau D}$ is the translation operator as given by Eq. (2.34). Substituting Eq. (3.5) for $\hat{a}(\theta, \tau)$ and doing some straightforward simplifications results in

$$A[\varphi] = \frac{1}{2\pi} \iint a\left(\tfrac{1}{2}(x+\tau), p\right) e^{i(x-\tau)p} \varphi(\tau) \, d\tau \, dp. \tag{3.16}$$

Also,

$$A[\varphi] = \frac{1}{2\pi} \iint a\left(x + \tfrac{1}{2}\tau, p\right) e^{-i\tau p} \varphi(\tau + x) \, d\tau \, dp. \tag{3.17}$$

3.3 Operation in the Fourier Domain

We now operate on functions in the Fourier domain. Using Eq. (3.10) we have

$$A[\hat{\varphi}] = A(X, D) \, \hat{\varphi}(p) = \iint \hat{a}(\theta, \tau) e^{-i\theta\tau/2} e^{i\tau p} e^{i\theta X} \hat{\varphi}(p) \, d\theta \, d\tau \tag{3.18}$$

$$= \iint \hat{a}(\theta, \tau) e^{-i\theta\tau/2} e^{i\tau p} \hat{\varphi}(p - \theta) \, d\theta \, d\tau. \tag{3.19}$$

Straightforward substitutions give

$$A[\hat{\varphi}] = \frac{1}{2\pi} \iint a(x, \tfrac{1}{2}(p+\theta)) \, e^{i(\theta-p)x} \, \hat{\varphi}(\theta) \, dx \, d\theta \tag{3.20}$$

and

$$A[\hat{\varphi}] = \frac{1}{2\pi} \iint a(x, p + \tfrac{1}{2}\theta) \, e^{i(\theta-p)x} \, \hat{\varphi}(\theta + p) \, dx \, d\theta. \tag{3.21}$$

That Eq. (3.20) and (3.16) are consistent with each other can be readily proven.

3.4 Operational Form

We now discuss a powerful procedure for calculating the Weyl operator. First, we describe a rearrangement procedure for operators that allows one to manipulate operators as ordinary functions. For any operator $A(X, D)$ define $R(x, p)$ by

$$R(x, p) = \text{rearrange } A(X, D), \text{ so that all the } D \text{ operators}$$
$$\text{are to the right of the } X \text{ operators; then replace } X, D \text{ by } x, p. \tag{3.22}$$

The rearrangement is achieved by using $[X, D] = i$. We point out that $R(x, p)$ can be the sum of rearranged factors. The concept of replacing a function of operators by a function of ordinary variables is a powerful way to manipulate operators and will often be used. After one manipulates $R(x, p)$ it must be then converted back to an operator and that is done by placing all the x factors to the left of the p factors and then substituting X, D for x, p. The ideas and methods were devised by Feynman in 1948 although the first instance of these types of calculations were introduced by McCoy in 1932. Feynman developed a general algebra for such manipulations.

In Eq. (3.9), $e^{i\theta X} e^{i\tau D}$ is already in the form where the D factors are to the right of the X factors and therefore

$$R(x,p) = \iint \hat{a}(\theta,\tau) \, e^{i\theta\tau/2} \, e^{i\theta x} \, e^{i\tau p} d\theta \, d\tau \tag{3.23}$$

$$= \iint \hat{a}(\theta,\tau) \, e^{\frac{1}{2i}\frac{\partial^2}{\partial x \partial p}} \, e^{i\theta x} \, e^{i\tau p} \, d\theta \, d\tau \tag{3.24}$$

$$= e^{\frac{1}{2i}\frac{\partial^2}{\partial x \partial p}} \iint \hat{a}(\theta,\tau) \, e^{i\theta x + i\tau p} \, d\theta \, d\tau. \tag{3.25}$$

Hence,

$$R(x,p) = \exp\left[\frac{1}{2i}\frac{\partial^2}{\partial x \partial p}\right] a(x,p). \tag{3.26}$$

The way Eq. (3.26) is used is to evaluate the right hand side and after the evaluation one puts all x factors to the left of the p factors and then substitutes X, D for x, p. Often the best way to evaluate $R(x,p)$ is to expand the exponential and then operate term by term. In particular,

$$R(x,p) = \sum_{n=0}^{\infty} \frac{1}{n!} \left(\frac{1}{2i}\right)^n \frac{\partial^{2n}}{\partial x^n \partial p^n} a(x,p). \tag{3.27}$$

Examples of these types of manipulations are given in the next section. Eq. (3.26) could have been obtained directly by way of Eq. (2.11), which we repeat here: for any two functions $f(\theta,\tau)$ and $\hat{a}(\theta,\tau)$,

$$\iint f(\theta,\tau) \hat{a}(\theta,\tau) \, e^{i\theta x + i\tau p} d\theta \, d\tau = f\left(\frac{1}{i}\frac{\partial}{\partial x}, \frac{1}{i}\frac{\partial}{\partial p}\right) a(x,p). \tag{3.28}$$

Taking $f(\theta,\tau) = e^{i\theta\tau/2}$ in Eq. (3.28) we immediately obtain Eq. (3.26).

3.5 Examples

We now give a number of examples chosen to illustrate the methods and to give insight into the nature of the Weyl operator.

Example. $a(x,p) = px$. Calculating the Fourier transform of px as per Eq. (3.5) we have

$$\hat{a}(\theta,\tau) = \frac{1}{4\pi^2} \iint xp \, e^{-i\theta x - i\tau p} \, dx \, dp = -\frac{\partial^2}{\partial \theta \partial \tau} \delta(\theta)\delta(\tau). \tag{3.29}$$

3.5. Examples

The Weyl operator, Eq. (3.9), is then

$$A(X,D) = -\iint \left\{ \frac{\partial^2}{\partial\theta\partial\tau}\delta(\theta)\delta(\tau) \right\} e^{i\theta\tau/2} e^{i\theta X} e^{i\tau D} \, d\theta \, d\tau \qquad (3.30)$$

$$= -\frac{\partial^2}{\partial\theta\partial\tau} e^{i\theta\tau/2} e^{i\theta X} e^{i\tau D} \Big|_{\theta,\tau=0} \qquad (3.31)$$

giving

$$A(X,D) = XD - \tfrac{1}{2}i. \qquad (3.32)$$

Using the commutation relation $XD - DX = i$ results in

$$A(X,D) = \tfrac{1}{2}(XD + DX). \qquad (3.33)$$

Therefore, for the symbol xp, the Weyl operator symmetrizes XD. However, this is not generally true for other symbols.

Alternatively, consider using Eq. (3.26),

$$R(x,p) = \exp\left[\frac{1}{2i}\frac{\partial^2}{\partial x \partial p}\right] a(x,p) = \left(1 + \frac{1}{2i}\frac{\partial^2}{\partial x \partial p} \cdots\right) xp = xp - \tfrac{1}{2}i. \qquad (3.34)$$

Now, we put the x factors before the p factors and then substitute X, D to obtain

$$A(X,D) = XD - \tfrac{1}{2}i \qquad (3.35)$$

which is the same as Eq. (3.32).

Example. $a(x,p) = x^2 p$. The Fourier transform is

$$\hat{a}(\theta,\tau) = \frac{1}{4\pi^2}\iint x^2 p\, e^{-i\theta x - i\tau p}\, dx\, dp = -i\frac{\partial^3}{\partial^2\theta\partial\tau}\delta(\theta)\delta(\tau) \qquad (3.36)$$

and therefore the Weyl operator is given by

$$A(X,D) = \iint \left\{ -i\frac{\partial^3}{\partial^2\theta\partial\tau}\delta(\theta)\delta(\tau) \right\} e^{i\theta\tau/2} e^{i\theta X} e^{i\tau D} \, d\theta \, d\tau \qquad (3.37)$$

$$= i\frac{\partial^3}{\partial^2\theta\partial\tau} e^{i\theta\tau/2} e^{i\theta X} e^{i\tau D} \Big|_{\theta,\tau=0} \qquad (3.38)$$

giving

$$A(X,D) = X^2 D - iX. \qquad (3.39)$$

We can obtain the same result using Eq. (3.27). In particular,

$$R(x,p) = \left(1 + \frac{1}{2i}\frac{\partial^2}{\partial x \partial p} - \frac{1}{4}\frac{\partial^4}{\partial x^2 \partial p^2} \cdots\right) x^2 p = x^2 p - ix \qquad (3.40)$$

which gives the same operator as Eq. (3.39) when (X,D) is substituted for (x,p). Clearly for this case using Eq. (3.40) is simpler.

Example. $a(x,p) = x^n p^m$. To obtain the Weyl operator for

$$a(x,p) = x^n p^m \qquad (3.41)$$

we use the method given by Eq. (3.27). We have

$$R(x,p) = \sum_{k=0}^{\infty} \frac{1}{k!} \left(\frac{1}{2i}\right)^k \frac{\partial^{2k}}{\partial x^k \partial p^k} x^n p^m. \qquad (3.42)$$

Using

$$\frac{\partial^k}{\partial x^k} x^n = \begin{cases} k!\binom{n}{k} x^{n-k} & k \leq n \\ 0 & k > n \end{cases} \qquad (3.43)$$

we obtain

$$R(x,p) = \sum_{k=0}^{\min(n,m)} \left(\frac{1}{2i}\right)^k k! \binom{n}{k}\binom{m}{k} x^{n-k} p^{m-k} \qquad (3.44)$$

and therefore

$$x^n p^m \leftrightarrow \sum_{k=0}^{\min(n,m)} \left(\frac{1}{2i}\right)^k k! \binom{n}{k}\binom{m}{k} X^{n-k} D^{m-k}. \qquad (3.45)$$

Similarly one can also show that

$$x^n p^m \leftrightarrow \sum_{k=0}^{\min(n,m)} \left(\frac{i}{2}\right)^k k! \binom{n}{k}\binom{m}{k} D^{m-k} X^{n-k}. \qquad (3.46)$$

Using Eq. (2.28) and (2.29) can be converted into

$$x^n p^m \leftrightarrow \frac{1}{2^n} \sum_{l=0}^{n} \binom{n}{l} X^{n-l} D^m X^l \qquad (3.47)$$

and

$$x^n p^m \leftrightarrow \frac{1}{2^m} \sum_{l=0}^{m} \binom{m}{l} D^{m-l} X^n D^l. \qquad (3.48)$$

These results were first derived by McCoy in 1932 [55].

Example. $a(x,p) = \delta(x-x')\delta(p-p')$. The delta function association will play an important role in generalizing the Weyl transform. We symbolize the association by

$$A_\delta(X,D) \leftrightarrow \delta(x-x')\delta(p-p'). \qquad (3.49)$$

3.5. Examples

The Fourier transform of $\delta(x-x')\delta(p-p')$ is

$$\hat{a}(\theta,\tau) = \frac{1}{4\pi^2}\iint \delta(x-x')\delta(p-p')\,e^{-i\theta x-i\tau p}\,dx\,dp = \frac{1}{4\pi^2}e^{-i\theta x'-i\tau p'} \quad (3.50)$$

and using Eq. (3.9), the Weyl operator is

$$A_\delta(X,D) = \iint \hat{a}(\theta,\tau)\,e^{i\theta\tau/2}\,e^{i\theta X}\,e^{i\tau D}\,d\theta\,d\tau \quad (3.51)$$

$$= \frac{1}{4\pi^2}\iint e^{-i\theta x'-i\tau p'}\,e^{i\theta\tau/2}\,e^{i\theta X}\,e^{i\tau D}\,d\theta\,d\tau \quad (3.52)$$

giving

$$A_\delta(X,D) = \frac{1}{2\pi}\int \delta(\tfrac{1}{2}\tau + X - x')\,e^{-i\tau p'}\,e^{i\tau D}\,d\tau. \quad (3.53)$$

Since the X factors are to the left of D factors we can immediately evaluate Eq. (3.53) to obtain $R_\delta(X,D)$:

$$R_\delta(x,p) = \frac{1}{\pi}e^{-2i(x'-x)(p'-p)}. \quad (3.54)$$

The operation on a function, $\varphi(x)$, is

$$A_\delta(X,D)\,\varphi(x) = \frac{1}{2\pi}\int \delta(\tfrac{1}{2}\tau + x - x')\,e^{-i\tau p'}\,\varphi(x+\tau)\,d\tau \quad (3.55)$$

$$= \frac{1}{\pi}e^{-2i(x'-x)p'}\,\varphi(2x'-x). \quad (3.56)$$

We can use the delta function association to define the operator for an arbitrary symbol. We first write

$$a(x,p) = \iint \delta(x-x')\delta(p-p')a(x',p')\,dx'\,dp' \quad (3.57)$$

and then in general

$$A(X,D)\,\varphi(x) = \iint A_\delta(X,D)\,\varphi(x)\,dx'\,dp'. \quad (3.58)$$

To show the consistency with our previous result substitute Eq. (3.56) into Eq. (2.37) to obtain

$$A(X,D)\,\varphi(x) = \frac{1}{\pi}\iint e^{-2i(x'-x)p'}\,\varphi(2x'-x)a(x',p')\,dx'\,dp'. \quad (3.59)$$

Making the transformation $\tau = 2x' - x$ we obtain

$$A(X,D)\,\varphi(x) = \frac{1}{2\pi}\iint a\left(\tfrac{1}{2}(x+\tau),p'\right)e^{ip'(x-\tau)}\,\varphi(\tau)\,d\tau\,dp' \quad (3.60)$$

which is the same as Eq. (3.16). Thus the delta function association fixes the association for a general symbol.

3.6 Inversion: From Weyl Operator to Symbol

We now address the question as to how to obtain the symbol from the Weyl operator. Given an operator, $A(X,D)$, we rearrange it by the procedure described in Sec. 3.4 to get $R(x,p)$. In particular, using Eq. (3.23), we have

$$R(x,p) = \iint \hat{a}(\theta,\tau) e^{i\theta\tau/2} e^{i\theta x} e^{i\tau p} \, d\theta \, d\tau. \tag{3.61}$$

Taking the inverse Fourier transform we obtain,

$$\hat{a}(\theta,\tau) e^{i\theta\tau/2} = \frac{1}{4\pi^2} \iint R(x,p) e^{-i\theta x - i\tau p} \, dx \, dp \tag{3.62}$$

or

$$\hat{a}(\theta,\tau) = \frac{1}{4\pi^2} e^{-i\theta\tau/2} \iint R(x,p) e^{-i\theta x - i\tau p} \, dx \, dp. \tag{3.63}$$

This shows how to get the Fourier transform of the symbol from $R(x,p)$. To express it in terms of the symbol directly we have

$$a(x,p) = \frac{1}{4\pi^2} \iint R(x',p') e^{-i\theta x' - i\tau p'} e^{-i\theta\tau/2} e^{i\theta x + i\tau p} \, d\theta \, d\tau \, dx' \, dp' \tag{3.64}$$

which evaluates to

$$a(x,p) = \frac{1}{\pi} \iint R(x',p') e^{2i(x'-x)(p'-p)} \, dx' \, dp' \tag{3.65}$$

or

$$a(x,p) = \frac{1}{\pi} \iint R(x'+x, p'+p) e^{2ix'p'} \, dx' \, dp'. \tag{3.66}$$

An alternative form can be obtained by just solving Eq. (3.26) for $a(x,p)$,

$$a(x,p) = \exp\left[-\frac{1}{2i}\frac{\partial}{\partial x}\frac{\partial}{\partial p}\right] R(x,p). \tag{3.67}$$

That Eq. (3.66) and (3.67) are indeed the same is a special case of Eq. (2.80).

Example. $A(X,D) = \frac{1}{2}(XD + DX)$. We first rearrange $A(X,D)$ so that all the X factors are to the left of the D factors. That results in

$$R(x,p) = xp - \tfrac{1}{2}i. \tag{3.68}$$

Then, using Eq. (3.16), we have

$$a(x,p) = \exp\left[-\frac{1}{2i}\frac{\partial}{\partial x}\frac{\partial}{\partial p}\right](xp - \tfrac{1}{2}i) \tag{3.69}$$

$$= \left(1 - \frac{1}{2i}\frac{\partial^2}{\partial x \partial p} - \frac{1}{4}\frac{\partial^4}{\partial x^2 \partial p^2}\right)(xp - \tfrac{1}{2}i) = xp \tag{3.70}$$

3.6. Inversion: From Weyl Operator to Symbol

which is consistent with the first example of Sect. 3.5. Alternatively using Eq. (3.65) we have

$$a(x,p) = \frac{1}{\pi} \iint (x'p' - \tfrac{1}{2}i) \, e^{2i(x'-x)(p'-p)} \, dx' \, dp' \qquad (3.71)$$

$$= \frac{1}{2\pi} \iint (x'p' - i) e^{2i(x'-x)(p'-p)} \, dx' \, dp' \qquad (3.72)$$

which evaluates straightforwardly to xp.

Example. $A(X,D) = D^2 X D$. Using the commutation relation to rearrange $D^2 X D$ so that all the x factors are to the left of the D factors we have

$$A(X,D) = D^2 X D = D(XD - i)D = DXD^2 - iD^2 \qquad (3.73)$$
$$= (XD - i)D^2 - iD^2 = XD^3 - 2iD^2. \qquad (3.74)$$

Therefore

$$R(x,p) = xp^3 - 2ip^2. \qquad (3.75)$$

Using Eq. (3.67) we obtain

$$a(x,p) = \exp\left[-\frac{1}{2i} \frac{\partial}{\partial x} \frac{\partial}{\partial p}\right] (xp^3 - 2ip^2) \qquad (3.76)$$

$$= xp^3 - \tfrac{1}{2}ip^2. \qquad (3.77)$$

Alternatively one can use Eq. (3.66),

$$a(x,p) = \frac{1}{\pi} \iint [(x' + x)(p' + p)^3 - 2i(p' + p)^2] \, e^{i2x'p'} \, dx' \, dp' \qquad (3.78)$$

$$= xp^3 - 2ip^2 + \frac{1}{\pi} \iint x'(p' + p)^3 \, e^{i2x'p'} \, dx' \, dp'. \qquad (3.79)$$

Straightforward evaluation of the last integral gives

$$\frac{1}{\pi} \iint x'(p' + p)^3 \, e^{i2x'p'} \, dx' \, dp' = -\tfrac{3}{2i}p^2 \qquad (3.80)$$

and therefore

$$a(x,p) = xp^3 - 2ip^2 - \tfrac{3}{2i}p^2 = xp^3 - \tfrac{1}{2}ip^2 \qquad (3.81)$$

which agrees with Eq. (3.77)

Alternate Approach. We now give another viewpoint for getting an inversion formula. Kim and Scully [42] argue that if we operate on a complete set of functions that should fix the operator and hence one should be able to obtain the corresponding symbol. The simplest complete set is the delta function and hence we consider, in Eq. (3.16), the case

$$\varphi(x) = \delta(x - x'') \tag{3.82}$$

giving

$$A[\delta(x - x'')] = \frac{1}{2\pi} \iint a\left(\tfrac{1}{2}(x+\tau), p\right) e^{i(x-\tau)p} \, \delta(\tau - x'') \, d\tau \, dp \tag{3.83}$$

$$= \frac{1}{2\pi} \int a\left(\tfrac{1}{2}(x+x''), p\right) e^{i(x-x'')p} \, dp. \tag{3.84}$$

Knowing $A[\delta(x - x'')]$ for all x'', one should be able to obtain the symbol, $a(x, p)$. Multiply Eq. (3.84) by $e^{-ip''x}$ and integrate with respect to x:

$$\int A[\delta(x - x'')] e^{-ip''x} dx = \frac{1}{2\pi} \iint a\left(\tfrac{1}{2}(x+x''), p\right) e^{-ip''x} e^{i(x-x'')p} \, dp \, dx \tag{3.85}$$

$$= \frac{1}{\pi} \iint a(x, p) \, e^{-i(2x-x'')p''} e^{2i(x-x'')p} \, dp \, dx. \tag{3.86}$$

Now multiply by $e^{ip''x'} e^{-2i(x''-x')(p''-p')}$ and integrate with respect to $dx''dp''$:

$$\int A[\delta(x-x'')] e^{-ip''x} e^{ip''x'} e^{-2i(x''-x')(p''-p')} \, dx \, dx'' \, dp'' \tag{3.87}$$

$$= \frac{1}{\pi} \iint \iint a(x,p) \, e^{ip''x'} e^{-2i(x''-x')(p''-p')} e^{-ip''(2x-x'')} e^{2i(x-x'')p} \, dp \, dx \, dx'' \, dp''.$$

Straightforward evaluation of the right hand side of Eq. (3.87) leads to $\pi a(x', p')$. Therefore

$$a(x', p') = \frac{1}{\pi} \iiint A[\delta(x-x'')] e^{-ip''x} e^{ip''x'} e^{-2i(x''-x')(p''-p')} \, dx \, dx'' \, dp''. \tag{3.88}$$

One can simplify further by integrating with respect to dp'' giving

$$a(x', p') = 2 \iiint A[\delta(x-x'')] \delta(x+x''-2x') e^{2i(x''-x')p'} \, dx \, dx'' \tag{3.89}$$

$$= 2 \iiint A[\delta(x-x'')] \delta(x+x''-2x') e^{2i(x'-x)p'} \, dx \, dx''. \tag{3.90}$$

3.6. Inversion: From Weyl Operator to Symbol

Example. $A(X, D) = XD$.

$$a(x', p') = \frac{1}{\pi} \iiint (xD\delta(x - x''))e^{-ip''x}e^{ip''x''}e^{-2i(x''-x')(p''-p')}dxdx''dp'' \quad (3.91)$$

$$= \frac{1}{\pi i} \iiint x\delta'(x - x'')e^{-ip''x}e^{ip''x''}e^{-2i(x''-x')(p''-p')}dxdx''dp'' \quad (3.92)$$

$$= -\frac{1}{\pi i} \iiint \left(\frac{d}{dx}[xe^{-ip''x}]_{x=x''}\right)e^{ip''x''}e^{-2i(x''-x')(p''-p')}dxdx''dp'' \quad (3.93)$$

$$= \frac{1}{\pi} \iiint [p''x'' + i]e^{-2i(x''-x')(p''-p')}dx''dp''. \quad (3.94)$$

The two integrals involved evaluate to

$$\frac{1}{\pi} \iiint p''x'' e^{-2i(x''-x')(p''-p')}dx''dp'' = p'x' + \frac{1}{2i} \quad (3.95)$$

$$\frac{1}{\pi} \iiint ie^{-2i(x''-x')(p''-p')}dx''dp'' = i \quad (3.96)$$

and therefore

$$a(x', p') = p'x' + \frac{1}{2i} + i = p'x' - \frac{1}{2i}. \quad (3.97)$$

It is instructive to also derive the result by way of Eq. (3.89)

$$a(x', p') = \frac{2}{i} \iiint x[\delta'(x - x'')]\delta(x + x'' - 2x')e^{2i(x''-x')p'}dxdx'' \quad (3.98)$$

$$= -\frac{2}{i} \iiint \frac{d}{dx}x\delta(x + x'' - 2x')|_{x=x''}e^{2i(x''-x')p'}dxdx'' \quad (3.99)$$

$$= -\frac{2}{i} \iiint \left[x''\frac{d}{dx}\delta(x + x'' - 2x')|_{x=x''} + \frac{1}{2}\delta(x'' - x')\right]e^{2i(x''-x')p'}dxdx''. \quad (3.100)$$

We note that

$$\frac{d}{dx}\delta(x + x'' - 2x')|_{x=x''} = \frac{1}{4}\frac{d}{dx''}\delta(x'' - x') \quad (3.101)$$

and hence

$$a(x', p') = -\frac{2}{i} \iiint \left[\frac{1}{4}x''\frac{d}{dx''}\delta(x'' - x') + \frac{1}{2}\delta(x'' - x')\right]e^{2i(x''-x')p'}dxdx'' \quad (3.102)$$

which evaluates to $a(x', p')$ as given by Eq. (3.97).

It is of interest to show Eq. (3.101) as such expressions often occur. We have

$$\frac{d}{dx}\delta(x+x''-2x')|_{x=x''} = \frac{1}{2\pi}\frac{d}{dx}\int e^{(x+x''-2x')y}dy|_{x=x''} \tag{3.103}$$

$$= \frac{1}{2\pi}\int y e^{(x''+x''-2x')y}dy \tag{3.104}$$

$$= \frac{1}{2\pi}\frac{1}{2}\frac{d}{dx''}\int y e^{(2x''-2x')y}dy \tag{3.105}$$

$$= \frac{1}{2}\frac{d}{dx''}\delta(2x''-2x') \tag{3.106}$$

$$= \frac{1}{4}\frac{d}{dx''}\delta(x''-x'). \tag{3.107}$$

3.7 Adjoint

The Hermitian adjoint of an operator, designated by $A^\dagger(X, D)$, is defined as the operator that makes the following true for any two functions $\psi(x)$ and $\varphi(x)$:

$$\int \psi^*(x) A(X, D) \varphi(x)\, dx = \int \varphi(x) \left\{ A^\dagger(x, D)\psi(x) \right\}^* dx. \tag{3.108}$$

If we take the adjoint of A in Eq. (3.7) we have

$$A^\dagger(X, D) = \iint \left[\widehat{a}(\theta, \tau) e^{i\theta X + i\tau D} \right]^\dagger d\theta\, d\tau. \tag{3.109}$$

But by Eq. (2.60) we have

$$\left[\widehat{a}(\theta, \tau) e^{i\theta X + i\tau D} \right]^\dagger = \widehat{a}^*(\theta, \tau) e^{-i\theta X - i\tau D} \tag{3.110}$$

and therefore

$$A^\dagger(X, D) = \iint \widehat{a}^*(\theta, \tau) e^{-i\theta X - i\tau D} d\theta\, d\tau \tag{3.111}$$

and

$$A^\dagger(X, D) = \iint \widehat{a}^*(-\theta, -\tau) e^{i\theta X + i\tau D} d\theta\, d\tau. \tag{3.112}$$

3.8 Hermiticity

Of particular importance is the question as to when the Weyl operator is Hermitian. The reason that Hermitian operators are important is that their expectation value, as defined by

$$\langle A(X, D) \rangle = \int \varphi^*(x) A(X, D)\, \varphi(x)\, dx \tag{3.113}$$

is assured to be real. In addition, in quantum mechanics and time-frequency analysis operators represent physical quantities and the numerical values of the physical quantities are the eigenvalues of the operator. Hence, one has to have real eigenvalues since measurable quantities are real. The way to ensure that the eigenvalues of an operator are real is to insist that the operator be Hermitian.

An operator is Hermitian if its adjoint equals the operator itself. If we compare Eq. (3.112) to Eq. (3.7) we see that if we take $\hat{a}^*(-\theta, -\tau) = \hat{a}(\theta, \tau)$ then the two expressions are equal. But this is precisely the condition for the symbol to be real

$$\hat{a}^*(-\theta, -\tau) = \hat{a}(\theta, \tau) \qquad \text{for real } a(x, p) \qquad (3.114)$$

and so we have

$$A^\dagger(X, D) = A(X, D) \qquad \text{for real } a(x, p). \qquad (3.115)$$

Therefore for real symbols the Weyl operator is Hermitian.

It is of some interest to prove the Hermitian property directly. The definition of a Hermitian operator, $A(X, D)$, is that for any two functions, φ and ψ,

$$\int \psi^*(x) A(X, D) \varphi(x) \, dx = \int \varphi(x) \left\{ A(X, D) \psi(x) \right\}^* dx. \qquad (3.116)$$

Substituting Eq. (3.16) into the left hand side of Eq. (3.116) we obtain

$$\int \psi^*(x) A(X, D) \varphi(x) \, dx = \frac{1}{2\pi} \iiint \psi^*(x) a \left(\tfrac{1}{2}(x + \tau), p \right) e^{ip(x-\tau)} \varphi(\tau) \, d\tau \, dp \, dx. \qquad (3.117)$$

A few manipulations lead to

$$\int \psi^*(x) A(X, D) \varphi(x) \, dx$$
$$= \frac{1}{2\pi} \iiint \varphi(x) \left\{ a^* \left(\tfrac{1}{2}(x + \tau), p \right) e^{-ip(x-\tau)} \psi(\tau) \right\}^* d\tau \, dp \, dx. \qquad (3.118)$$

If the symbol is real then the right hand side is exactly equal to the right hand side of Eq. (3.116) and hence $A(X, D)$ is Hermitian.

3.9 The Algebra of the Weyl Operator

We now discuss some basic properties of the Weyl operator [91].

Linearity. The Weyl association is linear in the sense that if

$$a(x, p) \leftrightarrow A(X, D), \qquad (3.119)$$
$$b(x, p) \leftrightarrow B(X, D), \qquad (3.120)$$

then
$$a(x,p) + b(x,p) \leftrightarrow A(X,D) + B(X,D). \tag{3.121}$$

Unit Association. The Weyl association associates the unit operator with the number 1:
$$1 \leftrightarrow I \tag{3.122}$$
where I is the unit operator.

Translated symbol. It often arises that we have to calculate the Weyl operator corresponding to
$$a_t(x,p) = a(x - x', p - p') \tag{3.123}$$
where x' and p' are constants. We denote by A_t the Weyl operator that corresponds to a_t,
$$A_t(X,D) \leftrightarrow a_t(x,p). \tag{3.124}$$
The Fourier transform of $a_t(x,p)$ is given by
$$\widehat{a_t}(\theta,\tau) = \frac{1}{4\pi^2} \iint a(x-x', p-p') e^{-i\theta x - i\tau p}\, dx\, dp = e^{-i\theta x' - i\tau p'} \widehat{a}(\theta,\tau). \tag{3.125}$$

Substituting into Eq. (3.9) gives
$$A_t(X,D) = \iint e^{-i\theta x' - i\tau p'} \widehat{a}(\theta,\tau) e^{i\theta\tau/2} e^{i\theta X} e^{i\tau D}\, d\theta\, d\tau \tag{3.126}$$
$$= \iint \widehat{a}(\theta,\tau) e^{i\theta\tau/2} e^{i\theta(X-x')} e^{i\tau(D-p')}\, d\theta\, d\tau \tag{3.127}$$

and hence
$$A_t(X,D) = A(X - x', D - p'). \tag{3.128}$$

The Scaled Symbol. Suppose we consider the scaled symbol
$$a_S(x,p) = a(\eta x, \mu p) \tag{3.129}$$
where η and μ are positive constants and we wish to obtain the corresponding Weyl operator. We denote it by
$$A_S(X,D) \leftrightarrow a_S(x,p). \tag{3.130}$$
The Fourier transform of $a_S(x,p)$ is
$$\widehat{a_S}(\theta,\tau) = \frac{1}{4\pi^2} \iint a(\eta x, \mu p) e^{-i\theta x - i\tau p}\, dx\, dp \tag{3.131}$$
$$= \frac{1}{4\pi^2 \eta\mu} \iint a(x,p) e^{-i\theta x/\eta - i\tau p/\mu}\, dx\, dp \tag{3.132}$$

3.9. The Algebra of the Weyl Operator

and therefore

$$\widehat{a_S}(\theta, \tau) = \frac{1}{4\pi^2 \eta\mu} \widehat{a}(\theta/\eta, \tau/\mu). \tag{3.133}$$

Substituting this into Eq. (3.9) we have

$$A_S(X, D) = \frac{1}{4\pi^2 \eta\mu} \iint \widehat{a}(\theta/\eta, \tau/\mu) \, e^{i\theta\tau/2} \, e^{i\theta X} \, e^{i\tau D} \, d\theta \, d\tau \tag{3.134}$$

$$= \frac{1}{4\pi^2} \iint \widehat{a}(\theta, \tau) \, e^{i\eta\mu\theta\tau/2} \, e^{i\theta\eta X} \, e^{i\tau\mu D} \, d\theta \, d\tau. \tag{3.135}$$

If we take

$$\eta\mu = 1 \tag{3.136}$$

then

$$A_S(X, D) = \frac{1}{4\pi^2} \iint \widehat{a}(\theta, \tau) \, e^{i\theta\tau/2} \, e^{i\theta\eta X} \, e^{i\tau\mu D} d\theta \, d\tau = A(\eta X, D/\eta). \tag{3.137}$$

Complex Conjugate of the Symbol. Suppose we have the function $a(x, p)$ and the corresponding Weyl operator $A(X, D)$ and wish to obtain the Weyl operator for the complex conjugate of $a(x, p)$. We use the notation $A_*(X, D)$ for the operator that corresponds to a^*:

$$A_*(X, D) \leftrightarrow a^*(x, p). \tag{3.138}$$

The Fourier transform of a^* is given by

$$\widehat{a^*}(\theta, \tau) = \widehat{a}(-\theta, -\tau) \tag{3.139}$$

and therefore the Weyl operator of a^* is

$$A_*(X, D) = \iint \widehat{a}(-\theta, -\tau) \, e^{i\theta X + i\tau D} \, d\theta \, d\tau \tag{3.140}$$

or

$$A_*(X, D) = \iint \widehat{a}(\theta, \tau) \, e^{-i\theta X - i\tau D} \, d\theta \, d\tau = A(-X, -D). \tag{3.141}$$

Derivative of the Symbol. We now obtain the Weyl operator for the derivative of a symbol. We use the notation a_x for the derivative with respect to x,

$$a_x(x, p) = \frac{\partial}{\partial x} a(x, p) \tag{3.142}$$

and the associated operator is denoted by

$$A_x(X, D) \leftrightarrow a_x(x, p). \tag{3.143}$$

The Fourier transform of $a_x(x, p)$ is

$$\widehat{\frac{\partial}{\partial x}a}(\theta, \tau) = \widehat{a_x}(\theta, \tau) = \frac{1}{4\pi^2} \iint \left[\frac{\partial}{\partial x}a(x,p)\right] e^{-i\theta x - i\tau p}\, dx\, dp = i\theta\, \widehat{a}(\theta, \tau) \quad (3.144)$$

and therefore

$$A_x(X, D) = i \iint \theta \widehat{a}(\theta, \tau)\, e^{i\theta\tau/2}\, e^{i\theta X}\, e^{i\tau D}\, d\theta\, d\tau = \frac{\partial}{\partial X} A(X, D). \quad (3.145)$$

One must be careful with the meaning of $\frac{\partial}{\partial X} A(X, D)$: the order must be preserved in the differentiation.

Now consider the derivative with respect to p. Using similar notation as above and writing

$$A_p(X, D) \leftrightarrow a_p(x, p) \quad (3.146)$$

the Fourier transform of $a_p(x, p)$ is

$$\widehat{\frac{\partial}{\partial p}a}(\theta, \tau) = \widehat{a_p}(\theta, \tau) = \frac{1}{4\pi^2} \iint \left[\frac{\partial}{\partial p}a(x,p)\right] e^{-i\theta x - i\tau p}\, dx\, dp = i\tau \widehat{a}(\theta, \tau). \quad (3.147)$$

Hence,

$$A_p(X, D) = i \iint \tau \widehat{a}(\theta, \tau)\, e^{i\theta\tau/2}\, e^{i\theta x}\, e^{i\tau D}\, d\theta\, d\tau \quad (3.148)$$

giving

$$A_p(X, D) = \frac{\partial}{\partial D} A(X, D). \quad (3.149)$$

Derivative of a Function. We aim at expressing the Weyl symbol operating on the derivative of a function in terms of the Weyl symbol of the function. We have

$$A\left[\frac{\partial}{\partial x}\varphi(x)\right] = \frac{1}{2\pi} \iint a\left(\tfrac{1}{2}(x+\tau), p\right) e^{ip(x-\tau)} \frac{\partial}{\partial \tau}\varphi(\tau)\, d\tau\, dp. \quad (3.150)$$

An integration by parts leads to

$$A\left[\frac{\partial}{\partial x}\varphi(x)\right] = -\frac{1}{2\pi} \iint \left\{\frac{\partial}{\partial \tau} a\left(\tfrac{1}{2}(x+\tau), p\right)\right\} e^{ip(x-\tau)} \varphi(\tau)\, d\tau\, dp$$

$$+ \frac{1}{2\pi} \iint a\left(\tfrac{1}{2}(x+\tau), p\right) \left\{\frac{\partial}{\partial \tau} e^{ip(x-\tau)}\right\} \varphi(\tau)\, d\tau\, dp. \quad (3.151)$$

We can set

$$\frac{\partial}{\partial \tau} a\left(\tfrac{1}{2}(x+\tau), p\right) = \frac{\partial}{\partial x} a\left(\tfrac{1}{2}(x+\tau), p\right) \quad (3.152)$$

and we then see that

$$A\left[\frac{\partial}{\partial x}\varphi(x)\right] = -\tfrac{1}{2} A_{a_x}[\varphi] + ip A[\varphi] \quad (3.153)$$

3.10. The Wigner distribution

where

$$a_x(x,p) = \frac{\partial}{\partial x} a(x,p), \tag{3.154}$$

$$A_{a_x}[\varphi] = \frac{1}{2\pi} \iint \left\{ \frac{\partial}{\partial x} a\left(\tfrac{1}{2}(x+\tau), p\right) \right\} e^{ip(x-\tau)} \varphi(\tau)\, d\tau\, dp. \tag{3.155}$$

Therefore, the Weyl operator operating on the derivative of a function is minus one half of the Weyl operator of the derivative of the symbol on the function plus ip of the Weyl operator.

3.10 The Wigner distribution

In subsequent chapters we develop the relation between operator correspondences with so-called quasi-probability distributions. Here, we just show, by way of expectation values, the direct relation between the Weyl operator and the Wigner distribution. Consider a symbol $a(x,p)$ and the corresponding Weyl operator $A(X,D)$. Let us for the moment accept the idea that the expectation value of an operator in a state $\varphi(x)$ is given by the inner product of φ with $A(x,D)\varphi$,

$$\langle A(X,D) \rangle = \int \varphi^*(x) A(X,D) \varphi(x)\, dx. \tag{3.156}$$

Using Eq. (3.16) we have

$$\int \varphi^*(x) A(X,D) \varphi(x)\, dx = \frac{1}{2\pi} \iint \varphi^*(x) a\left(\tfrac{1}{2}(x+\tau), p\right) e^{ip(x-\tau)} \varphi(\tau)\, d\tau\, dp\, dx. \tag{3.157}$$

Making the transformation

$$x \to x - \tfrac{1}{2}\tau, \qquad \tau \to x + \tfrac{1}{2}\tau, \tag{3.158}$$

we obtain that

$$\int \varphi^*(x) A(X,D) \varphi(x)\, dx = \frac{1}{2\pi} \iiint \varphi^*(x - \tfrac{1}{2}\tau) a(x,p)\, e^{-ip\tau}\, \varphi(x + \tfrac{1}{2}\tau)\, d\tau\, dp\, dx. \tag{3.159}$$

If we define $W(x,p)$, the Wigner distribution, by

$$W(x,p) = \frac{1}{2\pi} \int \varphi^*(x - \tfrac{1}{2}\tau)\, e^{-ip\tau}\, \varphi(x + \tfrac{1}{2}\tau)\, d\tau \tag{3.160}$$

we can write Eq. (3.159) as

$$\int \varphi^*(x) A(X,D) \varphi(x)\, dx = \iint a(x,p) W(x,p)\, dp\, dx. \tag{3.161}$$

It is important to appreciate that in Eq. (3.161), the left hand side deals with operators while the right hand side deals with ordinary functions only. Moreover, the right hand side is reminiscent of the expectation value of a (x,p) taken with the probability density $W(x,p)$. This is a very important interpretation and was the original motivation of Wigner [93]. As we discuss in later chapters $W(x,p)$ is not manifestly positive and hence can not be a proper probability density and that is why it is often called a quasi-probability distribution. Nonetheless Eq. (3.161) is the basis of what is commonly known as the phase-space representation of quantum mechanics and also plays a fundamental role in time-frequency analysis. The reason is that even though $W(x,p)$ is not a proper probability distribution it can be manipulated as if it was.

Fourier Domain. The expectation value of an operator can be evaluated in any domain and in particular for the Fourier domain we have

$$\int \varphi^*(x) A(x,D)\, \varphi(x)\, dx = \int \widehat{\varphi}^*(p) A(X,p)\, \widehat{u}(p)\, dp. \qquad (3.162)$$

Using Eq. (3.20) we write

$$\int \widehat{\varphi}^*(p) A(X,D)\, \widehat{\varphi}(p)\, dp = \frac{1}{2\pi} \iiint \widehat{\varphi}^*(p) a(x, \tfrac{1}{2}(p+\theta))\, e^{i(\theta-p)x}\, \widehat{\varphi}(\theta)\, dx\, d\theta\, dp. \qquad (3.163)$$

Making the transformation

$$p \to p + \tfrac{1}{2}\theta, \qquad \theta \to p - \tfrac{1}{2}\theta, \qquad (3.164)$$

we obtain

$$\int \widehat{\varphi}^*(p) A(X,p)\, \widehat{\varphi}(p)\, dp = \frac{1}{2\pi} \iint \widehat{\varphi}^*(p + \tfrac{1}{2}\theta) a(x,p) e^{-i\theta x}\, \widehat{\varphi}(p - \tfrac{1}{2}\theta)\, dx\, d\theta\, dp. \qquad (3.165)$$

which we write as

$$\int \widehat{\varphi}^*(p) A(X,D)\, \widehat{\varphi}(p)\, dp = \iint a(x,p)\, W(x,p)\, dx\, dp \qquad (3.166)$$

where now

$$W(x,p) = \frac{1}{2\pi} \int \widehat{\varphi}^*(p + \tfrac{1}{2}\theta) e^{-i\theta x}\, \widehat{\varphi}(p - \tfrac{1}{2}\theta)\, d\theta. \qquad (3.167)$$

That this expression of the Wigner distribution is the same as the one defined by Eq. (3.160) can be verified directly.

3.11 Product of Weyl Operators, Commutators, Moyal Bracket

The consideration of the product of two operators is important because the product will define the commutator of the operators and it is the commutator that enters in fundamental laws such as Heisenberg's equation of motion and the uncertainty principle. Suppose we have two symbols $a(x,p)$ and $b(x,p)$ with corresponding Weyl operators,

$$A(X,D) = \iint \hat{a}(\theta,\tau) e^{i\theta X + i\tau D} \, d\theta \, d\tau, \tag{3.168}$$

$$B(X,D) = \iint \hat{b}(\theta,\tau) e^{i\theta X + i\tau D} \, d\theta \, d\tau. \tag{3.169}$$

We consider the product of these two operators,

$$A(X,D)B(X,D) = \iiiint \hat{a}(\theta,\tau) e^{i\theta X + i\tau D} \hat{b}(\theta',\tau') e^{i\theta' X + i\tau' D} \, d\theta \, d\tau \, d\theta' \, d\tau'. \tag{3.170}$$

Note that the order of multiplication is important since, in general, $A(X,D)$ and $B(X,D)$ do not commute. We call the product $C(X,D)$,

$$C(X,D) = A(X,D)B(X,D). \tag{3.171}$$

We now ask, what is the symbol that corresponds to $C(X,D)$? That is, we seek the symbol $c(x,p)$ associated with $C(X,D)$,

$$c(x,p) \leftrightarrow C(X,D). \tag{3.172}$$

We want the symbol, $c(x,p)$, that corresponds to $C(X,D)$ as given by

$$C(X,D) = \iint \hat{c}(\theta,\tau) e^{i\theta X + i\tau D} \, d\theta \, d\tau. \tag{3.173}$$

Using Eq. (2.57)

$$e^{i\theta X + i\tau D} e^{i\theta' X + i\tau' D} = e^{i(\theta'\tau - \theta\tau')/2} e^{i(\theta+\theta')X + i(\tau+\tau')D} \tag{3.174}$$

Eq. (3.170) becomes

$$C(X,D) = A(X,D)B(X,D) \tag{3.175}$$

$$= \iiiint \hat{a}(\theta,\tau)\hat{b}(\theta',\tau') e^{i(\theta'\tau - \theta\tau')/2} e^{i(\theta+\theta')X + i(\tau+\tau')D} \, d\theta \, d\tau \, d\theta' \, d\tau' \tag{3.176}$$

$$= \iiiint \hat{a}(\theta - \theta', \tau - \tau')\hat{b}(\theta',\tau') e^{i(\theta'\tau - \theta\tau')/2} e^{i\theta X + i\tau D} \, d\theta \, d\tau \, d\theta' \, d\tau'. \tag{3.177}$$

Comparing with the general form, Eq. (3.173) we see that

$$\widehat{c}(\theta,\tau) = \iint \widehat{a}(\theta-\theta',\tau-\tau')\widehat{b}(\theta',\tau')\,e^{i(\theta'\tau-\theta\tau')/2}\,d\theta'\,d\tau'. \tag{3.178}$$

Taking the inverse Fourier transform we obtain

$$c(x,p) = \iiiint \widehat{a}(\theta,\tau)\widehat{b}(\theta',\tau')\,e^{i(\theta'\tau-\theta\tau')/2}e^{i(\theta+\theta')x+i(\tau+\tau')p}\,d\theta\,d\tau\,d\theta'\,d\tau'. \tag{3.179}$$

Substituting for $\widehat{a}(\theta,\tau)$ and $\widehat{b}(\theta',\tau')$ we have

$$c(x,p) = \left(\frac{1}{4\pi^2}\right)^2 \iint a(x',p')\,b(x'',p'')e^{i(\theta'\tau-\theta\tau')/2}e^{i(\theta+\theta')x+i(\tau+\tau')p}$$
$$\times e^{-i\theta x'-i\tau p'}\,e^{-i\theta' x''-i\tau' p''}\,dx'\,dx''\,dp'\,dp''\,d\theta\,d\tau\,d\theta'\,d\tau' \tag{3.180}$$

$$= \frac{1}{4\pi^2}\iint a(x',p')\,b(x'',p'')e^{i(\tau+\tau')p}e^{-i\tau p'}\,e^{-i\tau' p''}$$
$$\times \delta(\tau/2+x-x'')\delta(\tau'/2-x+x')\,dx'\,dx''\,dp'\,dp''\,d\tau\,d\tau' \tag{3.181}$$

which gives

$$c(x,p) = \frac{1}{\pi^2}\iiiint a(x',p')\,b(x'',p'')e^{2i[p(x''-x')+p'(x-x'')+p''(x'-x)]}\,dx'\,dx''\,dp'\,dp''. \tag{3.182}$$

By simple change of variables one can write a variety of forms for $c(x,p)$:

$$c(x,p) = \frac{1}{\pi^2}\iiiint a(x',p')\,b(x'',p'')e^{2i[x(p'-p'')+x'(p''-p)+x''(p-p')]}\,dx'\,dx''\,dp'\,dp'' \tag{3.183}$$

$$= \frac{1}{\pi^2}\iiiint a(x+x',p+p')\,b(x'',p'')e^{2i[x'(p''-p)-p'(x''-x)]}\,dx'\,dx''\,dp'\,dp'' \tag{3.184}$$

$$= \frac{1}{\pi^2}\iiiint a(x+x',p+p')\,b(x+x'',p+p'')e^{2i[x'p''-p'x'']}\,dx'\,dx''\,dp'\,dp''. \tag{3.185}$$

3.12 Operator Form

We now show interesting operator forms for $c(x,p)$,

$$c(x,p) = a\left(x+\frac{i}{2}\frac{\partial}{\partial p},p-\frac{i}{2}\frac{\partial}{\partial x}\right)b(x,p) \tag{3.186}$$

and

$$c(x,p) = \exp\left[\frac{i}{2}\left(\frac{\partial}{\partial x_a}\frac{\partial}{\partial p_b}-\frac{\partial}{\partial x_b}\frac{\partial}{\partial p_a}\right)\right]a(x,p)b(x,p). \tag{3.187}$$

3.12. Operator Form

In Eq. (3.187) the meaning of $\frac{\partial}{\partial x_a}$ is that it operates only on $a(x,p)$ and similarly for the other partial derivatives. To prove Eq. (3.186) we start with Eq. (3.179) and rewrite it as

$$c(x,p) = \iiiint \widehat{a}(\theta,\tau)\widehat{b}(\theta',\tau')\, e^{i\theta x + i\tau p}\, e^{i\theta'(x+\tau/2) + i\tau'(p-\theta/2)}\, d\theta\, d\tau\, d\theta'\, d\tau' \quad (3.188)$$

$$= \iiiint \widehat{a}(\theta,\tau)\, e^{i\theta x + i\tau p}\, \widehat{b}(\theta',\tau')\, e^{i\theta'(x+\tau/2) + i\tau'(p-\theta/2)}\, d\theta\, d\tau\, d\theta'\, d\tau' \quad (3.189)$$

$$= \iiiint \widehat{a}(\theta,\tau)\, e^{i\theta\left(x+\frac{i}{2}\frac{\partial}{\partial p}\right) + i\tau\left(p-\frac{i}{2}\frac{\partial}{\partial x}\right)}\, \widehat{b}(\theta',\tau')\, e^{i\theta' x + i\tau' p}\, d\theta\, d\tau\, d\theta'\, d\tau' \quad (3.190)$$

and considering $x + \frac{i}{2}\frac{\partial}{\partial p}$ and $p - \frac{i}{2}\frac{\partial}{\partial x}$ as if they were ordinary variables we have Eq. (3.186).

To prove Eq. (3.187) write Eq. (3.179) as

$$c(x,p) = \iiiint e^{i(\theta'\tau - \theta\tau')/2}\, \widehat{a}(\theta,\tau) e^{i\theta x + i\tau p}\, \widehat{b}(\theta',\tau') e^{i\theta' x + i\tau' p}\, d\theta\, d\tau\, d\theta'\, d\tau' \quad (3.191)$$

to obtain

$$c(x,p) = \iiiint e^{\frac{i}{2}\left(\frac{\partial}{\partial x_a}\frac{\partial}{\partial p_b} - \frac{\partial}{\partial x_b}\frac{\partial}{\partial p_a}\right)} \left[\widehat{a}(\theta,\tau) e^{i\theta x + i\tau p}\right]$$
$$\times \left[\widehat{b}(\theta',\tau') e^{i\theta' x + i\tau' p}\right] d\theta\, d\tau\, d\theta'\, d\tau' \quad (3.192)$$

and therefore Eq. (3.187) follows.

Star Notation. Common notation to indicate that $c(x,p)$ is composed from $a(x,p)$ and $b(x,p)$ by the procedure just described is the star notation. One writes

$$c_{ab}(x,p) = a(x,p) \star b(x,p). \quad (3.193)$$

Note that the order of $a(x,p)$ and $b(x,p)$ in $a(x,p) \star b(x,p)$ is important.

Commutators. We now obtain an expression for the symbol that corresponds to the commutator of $A(X,D)$ and $B(X,D)$ defined by

$$[A,B] = AB - BA. \quad (3.194)$$

The symbol corresponding to AB is

$$c_{ab}(x,p) = e^{\frac{i}{2}\left(\frac{\partial}{\partial x_a}\frac{\partial}{\partial p_b} - \frac{\partial}{\partial x_b}\frac{\partial}{\partial p_a}\right)} a(x,p) b(x,p) \quad (3.195)$$

and for BA we have

$$c_{ba}(x,p) = e^{-\frac{i}{2}\left(\frac{\partial}{\partial x_a}\frac{\partial}{\partial p_b} - \frac{\partial}{\partial x_b}\frac{\partial}{\partial p_a}\right)} a(x,p) b(x,p). \quad (3.196)$$

Hence for the commuter we have

$$[A, B] \leftrightarrow c_{ab}(x,p) - c_{ba}(x,p)$$
$$= \left[e^{\frac{i}{2}(\frac{\partial}{\partial x_a}\frac{\partial}{\partial p_b} - \frac{\partial}{\partial x_b}\frac{\partial}{\partial p_a})} - e^{-\frac{i}{2}(\frac{\partial}{\partial x_b}\frac{\partial}{\partial p_a} - \frac{\partial}{\partial x_a}\frac{\partial}{\partial p_b})} \right] a(x,p)b(x,p) \qquad (3.197)$$

giving

$$[A, B] \leftrightarrow 2i \sin \frac{1}{2} \left[\frac{\partial}{\partial x_a}\frac{\partial}{\partial p_b} - \frac{\partial}{\partial x_b}\frac{\partial}{\partial p_a} \right] a(x,p)b(x,p), \qquad (3.198)$$

a result first obtained by Moyal. The operator, $\sin \frac{1}{2} \left[\frac{\partial}{\partial x_a}\frac{\partial}{\partial p_b} - \frac{\partial}{\partial x_b}\frac{\partial}{\partial p_a} \right]$, is commonly called the Moyal bracket.

For the anti-commutator, using the same procedure, leads to

$$[A, B]_+ = AB + BA \leftrightarrow \left[e^{\frac{i}{2}(\frac{\partial}{\partial x_a}\frac{\partial}{\partial p_b} - \frac{\partial}{\partial x_b}\frac{\partial}{\partial p_a})} + e^{-\frac{i}{2}(\frac{\partial}{\partial x_b}\frac{\partial}{\partial p_a} - \frac{\partial}{\partial x_a}\frac{\partial}{\partial p_b})} \right] a(x,p)b(x,p)$$

giving

$$[A, B]_+ \leftrightarrow 2\cos \frac{1}{2} \left[\frac{\partial}{\partial x_a}\frac{\partial}{\partial p_b} - \frac{\partial}{\partial x_b}\frac{\partial}{\partial p_a} \right] a(x,p)b(x,p). \qquad (3.199)$$

Other forms. In addition, if we use Eq. (3.186) we have

$$[A, B] \leftrightarrow c_{ab}(x,p) - c_{ba}(x,p)$$
$$= a\left(x + \frac{i}{2}\frac{\partial}{\partial p}, p - \frac{i}{2}\frac{\partial}{\partial x} \right) b(x,p) - b\left(x + \frac{i}{2}\frac{\partial}{\partial p}, p - \frac{i}{2}\frac{\partial}{\partial x} \right) a(x,p)$$
$$(3.200)$$

and

$$[A, B]_+ = c_{ab}(x,p) + c_{ba}(x,p)$$
$$= a\left(x + \frac{i}{2}\frac{\partial}{\partial p}, p - \frac{i}{2}\frac{\partial}{\partial x} \right) b(x,p) + b\left(x + \frac{i}{2}\frac{\partial}{\partial p}, p - \frac{i}{2}\frac{\partial}{\partial x} \right) a(x,p).$$
$$(3.201)$$

Chapter 4

Generalized Operator Association

Besides the Weyl correspondence there are many other rules that have been proposed and we study them in Chapter 6. In this Chapter we develop a general formalism where all associations can be characterized in a simple way. This allows one to study all of them with a unified approach and to see the relationships between them [13, 48].

4.1 Generalized Operator

We define the generalized correspondence operator associated with the symbol $a(x,p)$ by

$$A^\Phi(X, D) = \iint \widehat{a}(\theta, \tau) \Phi(\theta, \tau) \, e^{i\theta X + i\tau D} \, d\theta \, d\tau \qquad (4.1)$$

where $\Phi(\theta, \tau)$ is a two-dimensional function called the kernel. The kernel characterizes a specific association and its properties. The general notation we use is that the capital letter in A^Φ signifies the operator corresponding to the lower case symbol $a(x,p)$ and the superscript Φ indicates the particular rule or kernel we are considering. The Weyl case is obtained for $\Phi = 1$.

For convenience we repeat here the definition of the Fourier transform of the symbol, $\widehat{a}(\theta, \tau)$,

$$\widehat{a}(\theta, \tau) = \frac{1}{4\pi^2} \iint a(x, p) \, e^{-i\theta x - i\tau p} \, dx \, dp \qquad (4.2)$$

and the symbol is given by $a(x,p)$:

$$a(x, p) = \iint \widehat{a}(\theta, \tau) \, e^{i\theta x + i\tau p} \, d\theta \, d\tau. \qquad (4.3)$$

Using Eq. (2.45),

$$e^{i\theta X+i\tau D} = e^{i\theta\tau/2} e^{i\theta X} e^{i\tau D} = e^{-i\theta\tau/2} e^{i\tau D} e^{i\theta X} \qquad (4.4)$$

we have that

$$A^{\Phi}(X,D) = \iint \hat{a}(\theta,\tau)\Phi(\theta,\tau) e^{i\theta\tau/2} e^{i\theta X} e^{i\tau D} \, d\theta \, d\tau \qquad (4.5)$$

$$= \iint \hat{a}(\theta,\tau)\Phi(\theta,\tau) e^{-i\theta\tau/2} e^{i\tau D} e^{i\theta X} \, d\theta \, d\tau. \qquad (4.6)$$

In terms of the symbol, one obtains

$$A^{\Phi}(X,D) = \frac{1}{4\pi^2} \iiiint a(q,p)\,\Phi(\theta,\tau)\, e^{-i\theta q - i\tau p + i\theta\tau/2}\, e^{i\theta X}\, e^{i\tau D}\, d\theta\, d\tau\, dq\, dp \qquad (4.7)$$

$$= \frac{1}{4\pi^2} \iiiint a(q,p)\,\Phi(\theta,\tau)\, e^{i\theta\tau/2}\, e^{i\theta(X-q)}\, e^{i\tau(D-p)}\, d\theta\, d\tau\, dq\, dp \qquad (4.8)$$

which may also be written as

$$A^{\Phi}(X,D) = \frac{1}{4\pi^2} \iiiint a(q,p)\,\Phi(\theta,\tau)\, e^{-i\theta q - i\tau p}\, e^{i\theta X + i\tau D}\, d\theta\, d\tau\, dq\, dp \qquad (4.9)$$

$$= \frac{1}{4\pi^2} \iiiint a(q,p)\,\Phi(\theta,\tau)\, e^{i\theta(X-q)+i\tau(D-p)}\, d\theta\, d\tau\, dq\, dp. \qquad (4.10)$$

4.2 Operational Form

For the operator $A^{\Phi}(X,D)$, we define R^{Φ} by the same procedure we used for the Weyl rule,

$$R^{\Phi}(x,p) = \text{rearrange } A^{\Phi}(X,D), \text{ so that all the } X \text{ factors are to the left of the } D \text{ operators; then replace } (X,D) \text{ by } (x,p). \qquad (4.11)$$

As with the Weyl case the rearrangement is achieved by using $[X,D] = i$. Applying this procedure to $A(X,D)$ as given by Eq. (4.5) and noting that the X operators are already to the left of the D factors we have

$$R^{\Phi}(x,p) = \iint \hat{a}(\theta,\tau)\Phi(\theta,\tau)\, e^{i\theta\tau/2} e^{i\theta x}\, e^{i\tau p}\, d\theta\, d\tau. \qquad (4.12)$$

Using Eq. (2.74), $R^{\Phi}(x,p)$ can be written in the following operator form,

$$R^{\Phi}(x,p) = \exp\left(\frac{1}{2i}\frac{\partial}{\partial x}\frac{\partial}{\partial p}\right) \Phi\left(\frac{1}{i}\frac{\partial}{\partial x}, \frac{1}{i}\frac{\partial}{\partial p}\right) a(x,p). \qquad (4.13)$$

4.3. The Operation on a Function

The way one uses Eq. (4.13) is to evaluate the right hand side first, then put all the x factors to the left of the p factors; then substitute (X, D) for (x, p). Since the Weyl case is when $\Phi = 1$ we have that

$$R^{\Phi}(x,p) = \Phi\left(\frac{1}{i}\frac{\partial}{\partial x}, \frac{1}{i}\frac{\partial}{\partial p}\right) R^{W}(x,p) \qquad (4.14)$$

where $R^W(x, p)$ is the rearrangement operator for the Weyl case.

4.3 The Operation on a Function

We now consider the operation of $A^{\Phi}(X, D)$ on an arbitrary function, $\varphi(x)$. Using Eq. (4.5) we have

$$A^{\Phi}[\varphi] = A^{\Phi}(X, D)\varphi(x) = \iint \hat{a}(\theta, \tau)\Phi(\theta, \tau)\, e^{i\theta\tau/2}\, e^{i\theta x}\, \varphi(x+\tau)\, d\theta\, d\tau \qquad (4.15)$$

where we have used the fact that $e^{i\tau D}\varphi(x) = \varphi(x+\tau)$. Equivalently,

$$A^{\Phi}[\varphi] = \iint \hat{a}(\theta, \tau - x)\Phi(\theta, \tau - x)\, e^{i\theta(x+\tau)/2}\, \varphi(\tau)\, d\theta\, d\tau. \qquad (4.16)$$

Writing $A^{\Phi}(X, D)$ in terms of the symbol one obtains

$$A^{\Phi}[\varphi] = \frac{1}{4\pi^2}\iiiint a\left(q + \frac{x+\tau}{2}, p\right) e^{-i\theta q + i(x-\tau)p}\, \Phi(\theta, \tau - x)\, \varphi(\tau)\, d\tau\, dq\, dp\, d\theta \qquad (4.17)$$

and,

$$A^{\Phi}[\varphi] = \frac{1}{4\pi^2}\iiiint a(q + x + \tau/2, p)\, e^{-i\theta q - i\tau p}\, \Phi(\theta, \tau)\, \varphi(\tau + x)\, d\tau\, dq\, dp\, d\theta. \qquad (4.18)$$

In addition,

$$A^{\Phi}[\varphi] = \frac{1}{4\pi^2}\iiiint a(q, p)\, e^{-i\theta(q - x - \tau/2) - i\tau p}\, \Phi(\theta, \tau)\, \varphi(\tau + x)\, d\tau\, dq\, dp\, d\theta. \qquad (4.19)$$

Notice that $\hat{a}(\theta, \tau - x)\Phi(\theta, \tau - x)\, e^{i\theta(\tau + x)/2}$ is a function of $\tau - x$ and $\tau + x$ and hence we may write

$$A^{\Phi}(X, D)\varphi(x) = \int k^{\Phi}(\tau + x, \tau - x)\, \varphi(\tau)\, d\tau \qquad (4.20)$$

with

$$k^{\Phi}(x, \tau) = \int \hat{a}(\theta, \tau)\Phi(\theta, \tau)\, e^{i\theta x/2}\, d\theta \qquad (4.21)$$

$$= \frac{1}{4\pi^2}\iiint a(q + \tfrac{1}{2}x, p)\, e^{-i\theta q - i\tau p}\Phi(\theta, \tau)\, dq\, dp\, d\theta. \qquad (4.22)$$

Fourier domain. For the operation on a function in the Fourier domain we use Eq. (4.6)

$$A^\Phi[\widehat{\varphi}] = A(X, D)\,\widehat{\varphi}(p) = \iint \widehat{a}(\theta, \tau)\Phi(\theta, \tau)e^{-i\theta\tau/2}\,e^{i\tau p}\,e^{i\theta X}\,\widehat{\varphi}(p)\,d\theta\,d\tau. \qquad (4.23)$$

Using

$$e^{i\theta X}\widehat{\varphi}(p) = \widehat{\varphi}(p - \theta) \qquad (4.24)$$

we have

$$A^\Phi[\widehat{\varphi}] = \iint \widehat{a}(\theta, \tau)\Phi(\theta, \tau)e^{-i\theta\tau/2}\,e^{i\tau p}\,\widehat{\varphi}(p - \theta)\,d\theta\,d\tau. \qquad (4.25)$$

In terms of the symbol, straightforward substitution gives that

$$A^\Phi[\widehat{\varphi}] = \frac{1}{4\pi^2}\iiiint a\left(x, p' + \tfrac{\theta - p}{2}\right)\Phi(p - \theta, \tau)e^{-i\tau p'}\,e^{i(\theta - p)x}\,\widehat{\varphi}(\theta)\,d\theta\,d\tau\,dx\,dp'. \qquad (4.26)$$

4.4 From Operator to Symbol

We now describe how starting with the operator $A^\Phi(X, D)$, one can obtain the corresponding symbol, $a(x, p)$. Inverting Eq. (4.12) we have,

$$\widehat{a}(\theta, \tau)\Phi(\theta, \tau)e^{i\theta\tau/2} = \frac{1}{4\pi^2}\iint R^\Phi(x, p)e^{-i\theta x - i\tau p}\,dx\,dp \qquad (4.27)$$

which gives

$$\widehat{a}(\theta, \tau) = \frac{1}{4\pi^2}\frac{e^{-i\theta\tau/2}}{\Phi(\theta, \tau)}\iint R^\Phi(x, p)e^{-i\theta x - i\tau p}\,dx\,dp \qquad (4.28)$$

and from which $a(x, p)$ can be obtained. Furthermore, from Eq. (4.13) we immediately have

$$a(x, p) = \exp\left[-\frac{1}{2i}\frac{\partial}{\partial x}\frac{\partial}{\partial p}\right]\Phi^{-1}\left(\frac{1}{i}\frac{\partial}{\partial x}, \frac{1}{i}\frac{\partial}{\partial p}\right)R^\Phi(x, p) \qquad (4.29)$$

$$= \frac{\exp\left[-\frac{1}{2i}\frac{\partial}{\partial x}\frac{\partial}{\partial p}\right]}{\Phi\left(\frac{1}{i}\frac{\partial}{\partial x}, \frac{1}{i}\frac{\partial}{\partial p}\right)}R^\Phi(x, p). \qquad (4.30)$$

4.5 Kernel From a Monomial Rule

We now consider how one can obtain the kernel for a particular correspondence rule that is given for the monomial $x^n p^m$. That is, suppose we are given

$$x^n p^m \leftrightarrow C_{nm}(X, D) \qquad (4.31)$$

4.5. Kernel From a Monomial Rule

where $C_{nm}(X, D)$ is the rule associated to the symbol $x^n p^m$. We wish to construct the corresponding kernel $\Phi(\theta, \tau)$. Often one can see it by inspection but sometimes not. For example, suppose we have the rule

$$x^n p^m \leftrightarrow X^n D^m \tag{4.32}$$

then one can guess that

$$e^{i\theta x + i\tau p} \leftrightarrow e^{i\theta X} e^{i\tau D} \tag{4.33}$$

and perhaps by comparing with Eq. (4.5) one can see that the kernel is given by

$$\Phi_S(\theta, \tau) = e^{-i\theta\tau/2}. \tag{4.34}$$

However suppose, for example, we have the rule proposed by Born-Jordan

$$x^n p^m \leftrightarrow \frac{1}{n+1} \sum_{\ell=0}^{m} X^{n-\ell} D^m X^\ell \tag{4.35}$$

then it is difficult to see what the kernel is.

We now give a general procedure and subsequently use it to calculate the Born-Jordan kernel. Consider the expansion of

$$e^{i\theta x + i\tau p} = \sum_{n=0}^{\infty} \sum_{m=0}^{\infty} \frac{(i\theta)^n (i\tau)^m}{n! m!} x^n p^m \tag{4.36}$$

and take the correspondence of each side

$$\Phi(\theta, \tau) e^{i\theta X + i\tau D} \leftrightarrow \sum_{n=0}^{\infty} \sum_{m=0}^{\infty} \frac{(i\theta)^n (i\tau)^m}{n! m!} C_{nm}(X, D). \tag{4.37}$$

Hence

$$\Phi(\theta, \tau) = e^{-i\theta X - i\tau D} \sum_{n=0}^{\infty} \sum_{m=0}^{\infty} \frac{(i\theta)^n (i\tau)^m}{n! m!} C_{nm}(X, D) \tag{4.38}$$

or

$$\Phi(\theta, \tau) = \sum_{n=0}^{\infty} \sum_{m=0}^{\infty} \frac{(i\theta)^n (i\tau)^m}{n! m!} e^{-i\theta X - i\tau D} C_{nm}(X, D). \tag{4.39}$$

In these equations it appears that the right hand side depends on X and D but in fact those factors will disappear once the calculation is explicitly done. Whether one uses Eq. (4.38) or Eq. (4.39) is a matter of convenience and taste.

Monomial from the kernel. We now consider how to obtain the monomial association if we are given the kernel. The right hand side of Eq. (4.37) is a Taylor series and hence

$$C_{nm}(X,D) = \frac{1}{i^n i^m} \frac{\partial^{n+m}}{\partial \theta^n \partial \tau^m} \Phi(\theta,\tau) e^{i\theta X + i\tau D}\Big|_{\theta,\tau=0} \quad (4.40)$$

$$= \frac{1}{i^n i^m} \frac{\partial^{n+m}}{\partial \theta^n \partial \tau^m} \Phi(\theta,\tau) e^{i\theta\tau/2} e^{i\theta X} e^{i\tau D}\Big|_{\theta,\tau=0} \quad (4.41)$$

$$= \frac{1}{i^n i^m} \frac{\partial^{n+m}}{\partial \theta^n \partial \tau^m} \Phi(\theta,\tau) e^{-i\theta\tau/2} e^{i\tau D} e^{i\theta X}\Big|_{\theta,\tau=0}. \quad (4.42)$$

Example. $a(x,p) = px$. Suppose one wants to characterize all correspondences that give

$$A^\Phi(X,D) = \tfrac{1}{2}(XD + DX) \leftrightarrow px. \quad (4.43)$$

One can show that would be the case if the kernel satisfies the properties

$$\Phi(0,0) = 1, \quad \frac{\partial}{\partial \theta}\Phi(\theta,\tau)\Big|_{\theta=0} = \frac{\partial}{\partial \tau}\Phi(\theta,\tau)\Big|_{\tau=0} = 0. \quad (4.44)$$

Examples are $\Phi(\theta,\tau) = \cos(\lambda\theta\tau)$ and $e^{-\lambda\theta^2\tau^2}$ where λ is a constant.

4.6 Algebra

The advantage of the kernel formulation is that one can readily obtain conditions on the kernel corresponding to properties one desires in the operator. We now discuss possible properties one may desire and the corresponding constraints on the kernel to ensure that requirement is met. Most of the results are easy to prove and we give them without proof.

Unit correspondence. If we want the correspondence between the number one and the unit operator I then we must take $\Phi(0,0) = 1$. That is

$$I \leftrightarrow 1 \quad \text{if} \quad \Phi(0,0) = 1. \quad (4.45)$$

Linearity. Linearity can be formulated as follows. If we have two symbols $a(x,p)$ and $b(x,p)$ corresponding to the operators $A^\Phi(X,D)$ and $B^\Phi(X,D)$,

$$A^\Phi(X,D) \leftrightarrow a(x,p), \quad (4.46)$$
$$B^\Phi(X,D) \leftrightarrow b(x,p), \quad (4.47)$$

then, we want

$$A^\Phi(X,D) + B^\Phi(X,D) \leftrightarrow a(x,p) + b(x,p). \quad (4.48)$$

This is assured for any kernel as long it is not a functional of $a(x,p)$ or $b(x,p)$.

4.7. Hermitian Adjoint

Symbols of x or p only. Suppose we want to be certain that for a symbol that is a function of x or p only, the corresponding operator is the same function of X and D. The respective conditions on the kernel are

$$a(x) \leftrightarrow a(X) \quad \text{if} \quad \Phi(0,\tau) = 1, \tag{4.49}$$
$$a(p) \leftrightarrow a(D) \quad \text{if} \quad \Phi(\theta,0) = 1. \tag{4.50}$$

Translation invariance. Consider the symbol $a_t(x,p) = a(x - x', p - p')$, then

$$\widehat{a}_t(\theta,\tau) = e^{-i\theta x' - i\tau p'} \widehat{a}(\theta,\tau) \tag{4.51}$$

and substituting into Eq. (4.1) one immediately has

$$A_t^\Phi(X,D) = A^\Phi(X - x', D - p'). \tag{4.52}$$

Eq. (4.52) is true for any kernel that does not explicitly depend on x or p.

4.7 Hermitian Adjoint

We designate the Hermitian adjoint by $A^{\Phi\dagger}(X,D)$. Since we know that the adjoint of $e^{i\theta X + i\tau D}$ is $e^{-i\theta X - i\tau D}$ we then have that

$$\left(\widehat{a}(\theta,\tau)\Phi(\theta,\tau)e^{i\theta X + i\tau D}\right)^\dagger = \widehat{a}^*(\theta,\tau)\Phi^*(\theta,\tau)e^{-i\theta X - i\tau D} \tag{4.53}$$

and therefore

$$A^{\Phi\dagger}(X,D) = \iint \widehat{a}^*(\theta,\tau)\Phi^*(\theta,\tau)\, e^{-i\theta X - i\tau D}\, d\theta\, d\tau \tag{4.54}$$

$$= \iint \widehat{a}^*(-\theta,-\tau)\Phi^*(-\theta,-\tau)\, e^{i\theta X + i\tau D}\, d\theta\, d\tau. \tag{4.55}$$

If the symbol is real then

$$\widehat{a}^*(-\theta,-\tau) = \widehat{a}(\theta,\tau) \qquad \text{for real symbols} \tag{4.56}$$

and if we further take

$$\Phi^*(-\theta,-\tau) = \Phi(\theta,\tau) \tag{4.57}$$

then we see that

$$A^{\Phi\dagger}(X,D) = A^\Phi(X,D) \qquad \text{for real symbols and for } \Phi^*(-\theta,-\tau) = \Phi(\theta,\tau). \tag{4.58}$$

Therefore the generalized operator is Hermitian for real symbols if the kernel satisfies the condition indicated in Eq. (4.57).

4.8 Product of Operators

We now address the question of operator multiplication. Suppose we have two symbols $a(x,p)$ and $b(x,p)$ with corresponding operators,

$$A^\Phi(X,D) = \iint \hat{a}(\theta,\tau)\Phi(\theta,\tau)\, e^{i\theta X + i\tau D}\, d\theta\, d\tau, \tag{4.59}$$

$$B^\Phi(X,D) = \iint \hat{b}(\theta,\tau)\, \Phi(\theta,\tau)\, e^{i\theta X + i\tau D}\, d\theta\, d\tau. \tag{4.60}$$

We consider the product of these two operators

$$C^\Phi(X,D) = A^\Phi(X,D) B^\Phi(X,D). \tag{4.61}$$

Explicitly,

$$A^\Phi(X,D)B^\Phi(X,D) = \iiiint \hat{a}(\theta,\tau)\Phi(\theta,\tau) e^{i\theta X + i\tau D}\, \hat{b}(\theta',\tau')\Phi(\theta',\tau')$$
$$\times e^{i\theta' X + i\tau' D}\, d\theta\, d\tau\, d\theta'\, d\tau'. \tag{4.62}$$

We seek the symbol, $c(x,p)$, that corresponds to $C^\Phi(X,D)$

$$c(x,p) \leftrightarrow C^\Phi(X,D) \tag{4.63}$$

where

$$C^\Phi(X,D) = \iint \hat{c}(\theta,\tau)\, e^{i\theta X + i\tau D}\, d\theta\, d\tau. \tag{4.64}$$

We give the results without details as the derivations are similar to the Weyl case given in Chap. 3. Using Eq. (2.57)

$$e^{i\theta X + i\tau D} e^{i\theta' X + i\tau' D} = e^{i(\theta'\tau - \theta\tau')/2} e^{i(\theta+\theta')X + i(\tau+\tau')D}, \tag{4.65}$$

Eq. (4.62) becomes

$$C^\Phi(X,D) = A^\Phi(X,D)B^\Phi(X,D) \tag{4.66}$$
$$= \iiiint \hat{a}(\theta-\theta', \tau-\tau')\Phi(\theta-\theta', \tau-\tau')\hat{b}(\theta',\tau')\Phi(\theta',\tau')$$
$$\times e^{i(\theta'\tau - \theta\tau')/2} e^{i\theta X + i\tau D}\, d\theta\, d\tau\, d\theta'\, d\tau'. \tag{4.67}$$

Comparing with Eq. (4.1) we see that

$$\hat{c}(\theta,\tau) = \iiiint \frac{\Phi(\theta-\theta', \tau-\tau')\Phi(\theta',\tau')}{\Phi(\theta,\tau)} \hat{a}(\theta-\theta', \tau-\tau')\hat{b}(\theta',\tau')$$
$$\times e^{i(\theta'\tau - \theta\tau')/2}\, d\theta'\, d\tau' \tag{4.68}$$

4.9. Transformation Between Associations

Taking the inverse Fourier transform and after some straightforward manipulations we obtain various forms:

$$c(x,p) = \iiiint \frac{\Phi(\theta,\tau)\Phi(\theta',\tau')}{\Phi(\theta+\theta',\tau+\tau')} \widehat{a}(\theta,\tau)\widehat{b}(\theta',\tau')$$
$$\times e^{i(\theta'\tau-\theta\tau')/2} e^{i(\theta+\theta')x+i(\tau+\tau')p} \, d\theta \, d\tau \, d\theta' \, d\tau' \qquad (4.69)$$

$$= \iiiint \frac{\Phi(\theta-\theta',\tau-\tau')\Phi(\theta',\tau')}{\Phi(\theta,\tau)} \widehat{a}(\theta-\theta',\tau-\tau')\widehat{b}(\theta',\tau')$$
$$\times e^{i(\theta'\tau-\theta\tau')/2} e^{i\theta x + i\tau p} \, d\theta \, d\tau \, d\theta' \, d\tau'. \qquad (4.70)$$

Operator Form. We now give the operational form in analogy to Eq. (3.187) for the Weyl case. In particular,

$$c(x,p) = L(a,b) \exp\left[\frac{i}{2}\left(\frac{\partial}{\partial x_a}\frac{\partial}{\partial p_b} - \frac{\partial}{\partial x_b}\frac{\partial}{\partial p_a}\right)\right] a(x,p) b(x,p) \qquad (4.71)$$

where we have defined

$$L(a,b) = \frac{\Phi(\frac{1}{i}\frac{\partial}{\partial x_a},\frac{1}{i}\frac{\partial}{\partial p_a})\Phi(\frac{1}{i}\frac{\partial}{\partial x_b},\frac{1}{i}\frac{\partial}{\partial p_b})}{\Phi(\frac{1}{i}\frac{\partial}{\partial x_a}+\frac{1}{i}\frac{\partial}{\partial x_b},\frac{1}{i}\frac{\partial}{\partial p_a}+\frac{1}{i}\frac{\partial}{\partial p_b})}, \qquad (4.72)$$

and we note that

$$L(a,b) = L(b,a). \qquad (4.73)$$

Commutators. For the commutator and anti-commutator of $A(X,D)$ and $B(X,D)$ one obtains

$$[A,B] \leftrightarrow 2i\sin\frac{1}{2}\left[\frac{\partial}{\partial x_a}\frac{\partial}{\partial p_b} - \frac{\partial}{\partial x_b}\frac{\partial}{\partial p_a}\right] L(a,b) a(x,p) b(x,p), \qquad (4.74)$$

$$[A,B]_+ \leftrightarrow 2\cos\frac{1}{2}\left[\frac{\partial}{\partial x_a}\frac{\partial}{\partial p_b} - \frac{\partial}{\partial x_b}\frac{\partial}{\partial p_a}\right] L(a,b) a(x,p) b(x,p). \qquad (4.75)$$

4.9 Transformation Between Associations

Suppose we have two different associations characterized by kernels $\Phi_1(\theta,\tau)$ and $\Phi_2(\theta,\tau)$ corresponding to $R^{\Phi_1}(x,p)$ and $R^{\Phi_2}(x,p)$, then using Eq. (4.13) we have

$$R^{\Phi_1}(x,p) = \exp\left(\frac{1}{2i}\frac{\partial}{\partial x}\frac{\partial}{\partial p}\right) \Phi_1\left(\frac{1}{i}\frac{\partial}{\partial x},\frac{1}{i}\frac{\partial}{\partial p}\right) a(x,p), \qquad (4.76)$$

$$R^{\Phi_2}(x,p) = \exp\left(\frac{1}{2i}\frac{\partial}{\partial x}\frac{\partial}{\partial p}\right) \Phi_2\left(\frac{1}{i}\frac{\partial}{\partial x},\frac{1}{i}\frac{\partial}{\partial p}\right) a(x,p), \qquad (4.77)$$

from which it follows that

$$R^{\Phi_2}(x,p) = \frac{\Phi_2\left(\frac{1}{i}\frac{\partial}{\partial x}, \frac{1}{i}\frac{\partial}{\partial p}\right)}{\Phi_1\left(\frac{1}{i}\frac{\partial}{\partial x}, \frac{1}{i}\frac{\partial}{\partial p}\right)} R^{\Phi_1}(x,p). \tag{4.78}$$

Also one can express the transformation in terms of the $k(x,\tau)$ in Eq. (4.20)

$$A^{\Phi_1}[\varphi] = \int k^{\Phi_1}(x+\tau, \tau-x)\,\varphi(\tau)\,d\tau, \tag{4.79}$$

$$A^{\Phi_2}[\varphi] = \int k^{\Phi_2}(x+\tau, \tau-x)\,\varphi(\tau)\,d\tau. \tag{4.80}$$

Simple manipulation of Eq. (4.20) leads to

$$k^{\Phi_2}(x,\tau) = \frac{1}{2\pi}\iint \frac{\Phi_2(2\theta,\tau)}{\Phi_1(2\theta,\tau)} e^{i\theta(x-x')}\, k^{\Phi_1}(x',\tau)\,d\theta\,dx'. \tag{4.81}$$

4.10 The Fourier, Taylor, and Delta Function Associations

The kernel formulation can be viewed profitably from different perspectives. The general idea is that one can expand a symbol in different ways and each one can be used to obtain the operator corresponding to the symbol. In particular, we consider three expansions, the Fourier, Taylor (monomial) and delta function expansions which are given respectively by

$$a(x,p) = \begin{cases} \iint \hat{a}(\theta,\tau)\,e^{i\theta x+i\tau p}\,d\theta\,d\tau & \text{Fourier,} \\ \sum_{n,m=0}^{\infty} \frac{1}{n!m!}\left\{\frac{\partial^{n+m}}{\partial x^n \partial p^m}a(x,p)\big|_{x,p=0}\right\} x^n p^m & \text{Taylor series (monomial),} \\ \iint a(x',p')\delta(x-x')\delta(p-p')\,dx'\,dp' & \text{delta function.} \end{cases} \tag{4.82}$$

Since all of these associations are linear, if we know the association for $e^{i\theta x+i\tau p}$, or $x^n p^m$, or for $\delta(x-x')\delta(p-p')$ we will get the association for a general symbol. We write the association for each case as follows,

$$A_F(X,D) \leftrightarrow e^{i\theta x+i\tau p}, \tag{4.83}$$
$$A_T(X,D) \leftrightarrow x^n p^m, \tag{4.84}$$
$$A_\delta(X,D) \leftrightarrow \delta(x)\delta(p). \tag{4.85}$$

We now discuss each of the cases.

4.10. The Fourier, Taylor, and Delta Function Associations

Fourier Association. One thinks of $e^{i\theta x+i\tau p}$ as a symbol with parameters θ and τ and associates

$$A_F(X,D) = \Phi(\theta,\tau)\, e^{i\theta X+i\tau D} \leftrightarrow e^{i\theta x+i\tau p}. \tag{4.86}$$

Hence, one argues, for a general symbol, expand the symbol in terms of its Fourier transform

$$a(x,p) = \iint \hat{a}(\theta,\tau)\, e^{i\theta x+i\tau p}\, d\theta\, d\tau \tag{4.87}$$

and therefore

$$A(X,D) = \iint \hat{a}(\theta,\tau)\, A_F(X,D)\, d\theta\, d\tau \tag{4.88}$$

$$= \iint \Phi(\theta,\tau)\hat{a}(\theta,\tau)\, e^{i\theta x+i\tau p}\, d\theta\, d\tau \tag{4.89}$$

which is Eq. (4.1).

Taylor Series Association. Suppose there is an operator correspondence for $x^n p^m$ and denote it by $C_{nm}(X,D)$,

$$A_T(X,D) = C_{nm}(X,D) \leftrightarrow x^n p^m. \tag{4.90}$$

Expand the symbol in a Taylor series

$$a(x,p) = \sum_{n,m=0}^{\infty} \frac{1}{n!m!} \left\{ \frac{\partial^{n+m}}{\partial x^n \partial p^m} a(x,p)\bigg|_{x,p=0} \right\} x^n p^m \tag{4.91}$$

and define the operator by

$$A^{\Phi}(X,D) = \sum_{n,m=0}^{\infty} \frac{1}{n!m!} \left\{ \frac{\partial^{n+m}}{\partial x^n \partial p^m} a(x,p)\bigg|_{x,p=0} \right\} C_{nm}(X,D). \tag{4.92}$$

To make this $A^{\Phi}(X,D)$ equal to the form given by Eq. (4.1) one takes

$$C_{nm}(X,D) = \frac{1}{i^n i^m} \frac{\partial^{n+m}}{\partial \theta^n \partial \tau^m} \Phi(\theta,\tau)\, e^{i\theta X+i\tau D} \bigg|_{\theta,\tau=0}. \tag{4.93}$$

Delta Function Association. We write the correspondence between $\delta(x)\delta(p)$ and the corresponding operator by $A_\delta(x,D)$ as

$$A_\delta^{\Phi}(X,D) \leftrightarrow \delta(x)\delta(p). \tag{4.94}$$

Starting with the identity

$$a(x,p) = \iint a(x',p')\delta(x-x')\delta(p-p')\, dx'\, dp' \tag{4.95}$$

we then have

$$A^\Phi(X, D) = \iint a(x', p') A^\Phi_\delta(X - x', D - p') \, dx' dp'. \tag{4.96}$$

Comparing to Eq. (4.1) we see that we must take

$$A^\Phi_\delta(X, D) = \frac{1}{4\pi^2} \iint \Phi(\theta, \tau) e^{i\theta X + i\tau D} \, d\theta \, d\tau \leftrightarrow \delta(x)\delta(p). \tag{4.97}$$

4.11 The Form of the Generalized Correspondence

In our formulation we have taken the correspondence for the exponential to be

$$e^{i\theta x + i\tau p} \leftrightarrow \Phi(\theta, \tau) e^{i\theta X + i\tau D} \tag{4.98}$$

and since $e^{i\theta X + i\tau D}$ is the correspondence for the Weyl symbol it may appear that we have elevated the Weyl association to a special status. However we could have used any correspondence rule as the basic rule. For example suppose we had defined

$$A^{\Phi'}(X, D) = \iint \hat{a}(\theta, \tau) \Phi'(\theta, \tau) e^{i\theta X} e^{i\tau D} d\theta \, d\tau \tag{4.99}$$

as the generalized correspondence rule where $\Phi'(\theta, \tau)$ is the kernel. Then for the Weyl case the kernel would not be one but it would have been $\Phi'(\theta, \tau) = e^{i\theta\tau/2}$.

4.12 The Kernel for the Born-Jordan Rule

In 1926 Born and Jordan [10] gave the following correspondence rule for a monomial, $x^n p^m$,

$$x^n p^m \leftrightarrow \frac{1}{n+1} \sum_{\ell=0}^{m} X^{n-\ell} D^m X^\ell. \tag{4.100}$$

The corresponding kernel was derived in reference [13]. Using Eq. (4.39) we have

$$\Phi^{BJ}(\theta, \tau) = \sum_{n=0}^{\infty} \sum_{m=0}^{\infty} \sum_{\ell=0}^{n} \frac{(i\theta)^n (i\tau)^m}{n! m! (n+1)} e^{-i\theta X - i\tau D} X^{n-\ell} D^m X^\ell. \tag{4.101}$$

The summation over m gives $e^{i\tau D}$ and therefore

$$\Phi^{BJ}(\theta, \tau) = \sum_{n=0}^{\infty} \sum_{\ell=0}^{n} \frac{(i\theta)^n}{n!(n+1)} e^{-i\theta X - i\tau D} X^{n-\ell} e^{i\tau D} X^\ell. \tag{4.102}$$

But

$$e^{-i\theta X - i\tau D} X^{n-\ell} e^{i\tau D} X^\ell = e^{i\theta\tau/2} e^{-i\theta x} e^{-i\tau D} x^{n-\ell} e^{i\tau D} x^\ell \tag{4.103}$$

$$= e^{i\theta\tau/2} e^{-i\theta x} (x - \tau)^{n-\ell} e^{i\tau D} (x - \tau)^\ell e^{-i\tau D} \tag{4.104}$$

$$= e^{i\theta\tau/2} e^{-i\theta x} (x - \tau)^{n-\ell} x^\ell \tag{4.105}$$

4.12. The Kernel for the Born-Jordan Rule

where in going from (4.104) to (4.105) we have used the fact that

$$e^{i\tau D}(X-\tau)^\ell e^{-i\tau D} = X^\ell. \tag{4.106}$$

Hence we have that

$$\Phi^{BJ}(\theta,\tau) = e^{i\theta\tau/2} e^{-i\theta x} \sum_{n=0}^{\infty} \sum_{\ell=0}^{n} \frac{(i\theta)^n}{n!(n+1)} (x-\tau)^{n-\ell} x^\ell \tag{4.107}$$

The remaining summations are elementary resulting in

$$\Phi^{BJ}(\theta,\tau) = \frac{\sin \frac{1}{2}\theta\tau}{\frac{1}{2}\theta\tau}. \tag{4.108}$$

Chapter 5

Generalized Phase-Space Distributions

Shortly after the invention of quantum mechanics, Wigner [93] and Kirkwood [43] addressed the issue of quantum statical mechanics in the following way. They devised a distribution function (different ones) aimed to calculate quantum averages by way of phase-space averaging. It was some time later that Moyal [60] saw the connection between the Weyl rule and the Wigner distribution. In this chapter, we develop a formulation where all distributions may be studied in a unified way. The basic result is that one can define an infinite number of distributions, $C(x,p)$, so that for a symbol $a(x,p)$ and state $\varphi(x)$,

$$\int \varphi^*(x)\, A^\Phi(X,D)\, \varphi(x)\, dx = \iint a(x,p)\, C(x,p)\, dx\, dp \tag{5.1}$$

where [13]

$$C(x,p) = \frac{1}{4\pi^2} \iiint \varphi^*(q-\tfrac{1}{2}\tau)\, \varphi(q+\tfrac{1}{2}\tau)\, \Phi(\theta,\tau) e^{-i\theta x - i\tau p + i\theta q}\, d\theta\, d\tau\, dq \tag{5.2}$$

and where $\Phi(\theta,\tau)$ is the kernel that was discussed in Chapter 4. We call $C(x,p)$ the generalized phase-space distribution. There are two ways to derive $C(x,p)$. We give the first here and the second in the next section.

For the generalized association

$$A^\Phi(X,D) = \iint \widehat{a}(\theta,\tau)\, \Phi(\theta,\tau)\, e^{i\theta X + i\tau D}\, d\theta\, d\tau \leftrightarrow a(x,p) \tag{5.3}$$

we showed in Eq. (4.19) that

$$A^\Phi[\varphi] = \frac{1}{4\pi^2} \iiiint a(q,p)\, \Phi(\theta,\tau) e^{-i\theta(q-x-\tau/2) - i\tau p}\, \varphi(\tau+x)\, d\tau\, dq\, dp\, d\theta. \tag{5.4}$$

Multiply by $\varphi^*(x)$ and integrate both sides to obtain

$$\int \varphi^*(x)\, A^\Phi\,(X,D)\, \varphi(x)\, dx = \frac{1}{4\pi^2} \int \cdots \int \varphi^*(x)\, a\,(q,p)\, e^{-i\theta(q-x-\tau/2)-i\tau p}$$
$$\times \Phi(\theta,\tau)\, \varphi(\tau+x)\, d\tau\, dq\, dp\, d\theta\, dx \tag{5.5}$$

or

$$\int \varphi^*(x)\, A^\Phi\,(X,D)\, \varphi(x)\, dx = \frac{1}{4\pi^2} \int \cdots \int \varphi^*(x - \tfrac{1}{2}\tau)\, a\,(q,p)$$
$$\times e^{-i\theta q + i\theta x - i\tau p}\, \Phi(\theta,\tau)\, \varphi(x + \tfrac{1}{2}\tau)\, d\tau\, dq\, dp\, d\theta\, dx. \tag{5.6}$$

For convenience interchange x and q on the right hand side,

$$\int \varphi^*(x)\, A^\Phi\,(X,D)\, \varphi(x)\, dx = \frac{1}{4\pi^2} \int \cdots \int \varphi^*(q - \tfrac{1}{2}\tau)\, a\,(x,p)$$
$$\times e^{-i\theta x + i\theta q - i\tau p}\, \Phi(\theta,\tau)\, \varphi(q + \tfrac{1}{2}\tau)\, d\tau\, dq\, dp\, d\theta\, dx. \tag{5.7}$$

Comparing to Eq. (5.1) we see that we must take

$$C(x,p) = \frac{1}{4\pi^2} \iiint \varphi^*(q - \tfrac{1}{2}\tau)\, \varphi(q + \tfrac{1}{2}\tau)\, \Phi(\theta,\tau)\, e^{-i\theta x - i\tau p + i\theta q}\, d\theta\, d\tau\, dq \tag{5.8}$$

which is the generalized phase-space distribution.

Fourier domain. Substituting the Fourier transform of the state function

$$\widehat{\varphi}(p) = \frac{1}{\sqrt{2\pi}} \int \varphi(x)\, e^{-ixp}\, dx \tag{5.9}$$

into Eq. (5.2) results in

$$C(x,p) = \frac{1}{4\pi^2} \iiint \Phi(\theta,\tau)\, \widehat{\varphi}^*(k + \tfrac{1}{2}\theta)\, e^{i\tau k - i\theta x - i\tau p}\, \widehat{\varphi}(k - \tfrac{1}{2}\theta)\, d\theta\, d\tau\, dk. \tag{5.10}$$

Cross distribution. One can readily prove that for two arbitrary functions $\psi(x)$ and $\varphi(x)$,

$$\int \psi^*(x)\, A\,(X,D)\, \varphi(x)\, dx = \iint a(x,p)\, C_{\psi\varphi}(x,p)\, dx\, dp \tag{5.11}$$

if we take

$$C_{\psi\varphi}(x,p) = \frac{1}{4\pi^2} \iiint \psi^*(q - \tfrac{1}{2}\tau)\, \varphi(q + \tfrac{1}{2}\tau)\, \Phi(\theta,\tau)\, e^{-i\theta x - i\tau p + i\theta q}\, d\theta\, d\tau\, dq. \tag{5.12}$$

5.1 Characteristic Function Approach

We now present an alternate derivation of the generalized distribution, Eq. (5.2). For two random variables, x and y, with probability distribution $P(x,y)$, the characteristic function, $M(\theta, \tau)$ is defined as the average of $e^{i\theta x + i\tau y}$,

$$M(\theta, \tau) = \langle e^{i\theta x + i\tau y} \rangle = \iint e^{i\theta x + i\tau y} P(x,y)\, dx\, dy \qquad (5.13)$$

and the distribution function is given by

$$P(x,y) = \frac{1}{4\pi^2} \iint M(\theta, \tau)\, e^{-i\theta x - i\tau y}\, d\theta\, d\tau. \qquad (5.14)$$

Further, by expanding the exponential in Eq. (5.13) one has

$$M(\theta, \tau) = \sum_{n=0}^{\infty} \sum_{m=0}^{\infty} \frac{(i\theta)^n (i\tau)^m}{n!\, m!} \langle x^n y^m \rangle \qquad (5.15)$$

where $\langle x^n y^m \rangle$ are the joint moments of the distribution

$$\langle x^n y^m \rangle = \iint x^n y^m\, P(x,y)\, dx\, dy. \qquad (5.16)$$

Considering $M(\theta, \tau)$ in Eq. (5.15) as a Taylor series we also have

$$\langle x^n y^m \rangle = \frac{1}{i^n i^m} \frac{\partial^{n+m}}{\partial \theta^n \partial \tau^m} M(\theta, \tau) \bigg|_{\theta, \tau = 0}. \qquad (5.17)$$

Now, the characteristic function is an average, the average of $e^{i\theta x + i\tau p}$, and hence using the association

$$e^{i\theta x + i\tau p} \leftrightarrow \Phi(\theta, \tau)\, e^{i\theta X + i\tau D} \qquad (5.18)$$

we have

$$M(\theta, \tau) = \langle \Phi(\theta, \tau)\, e^{i\theta X + i\tau D} \rangle = \Phi(\theta, \tau) \int \varphi^*(x)\, e^{i\theta X + i\tau D}\, \varphi(x)\, dx. \qquad (5.19)$$

Using Eq. (2.46) we have

$$e^{i\theta X + i\tau D}\, \varphi(x) = e^{i\theta \tau / 2 + i\theta x}\, \varphi(x + \tau) \qquad (5.20)$$

and substituting this into Eq. (5.19) one obtains

$$\int \varphi^*(x)\, e^{i\theta X + i\tau D}\, \varphi(x)\, dx = \int \varphi^*(x - \tfrac{1}{2}\tau)\, e^{i\theta x}\, \varphi(x + \tfrac{1}{2}\tau)\, dx \qquad (5.21)$$

and therefore

$$M(\theta,\tau) = \Phi(\theta,\tau) \int \varphi^*(x - \tfrac{1}{2}\tau) \, e^{i\theta x} \varphi(x + \tfrac{1}{2}\tau) \, dx. \tag{5.22}$$

Substituting this into Eq. (5.14) one obtains

$$C(x,p) = \frac{1}{4\pi^2} \iint M(\theta,\tau) \, e^{-i\theta x - i\tau p} \, d\theta \, d\tau \tag{5.23}$$

$$= \frac{1}{4\pi^2} \iiint \varphi^*(q - \tfrac{1}{2}\tau)\varphi(q + \tfrac{1}{2}\tau)\Phi(\theta,\tau) \, e^{-i\theta x - i\tau p + i\theta q} \, d\theta \, d\tau \, dq \tag{5.24}$$

which is the generalized distribution, Eq. (5.2).

5.2 Marginal Conditions

If we consider $C(x,p)$ a joint density, then integrating out one variable gives what are called the marginal densities, that is, the density of each variable. Integrating with respect to p and x one obtains that

$$\int C(x,p) \, dp = \frac{1}{2\pi} \iint \Phi(\theta,0) \, |\varphi(q)|^2 \, e^{i\theta(q-x)} \, d\theta \, dq, \tag{5.25}$$

$$\int C(x,p) \, dx = \frac{1}{2\pi} \iint \Phi(0,\tau) \, |\widehat{\varphi}(k)|^2 \, e^{i\tau(k-p)} \, d\tau \, dk. \tag{5.26}$$

If we want the marginals to be $|\varphi(x)|^2$ and $|\widehat{\varphi}(p)|^2$, then the constraints on the kernel are as indicated in the next two equations

$$\int C(x,p) \, dp = |\varphi(x)|^2 \quad \text{if} \quad \Phi(\theta,0) = 1, \tag{5.27}$$

$$\int C(x,p) \, dx = |\widehat{\varphi}(p)|^2 \quad \text{if} \quad \Phi(0,\tau) = 1. \tag{5.28}$$

5.3 Relation Between Distributions

Suppose we have two distributions, C_1 and C_2, with corresponding kernels, Φ_1 and Φ_2. From Eq. (5.22) their characteristic functions are

$$M_1(\theta,\tau) = \Phi_1(\theta,\tau) \int \varphi^*(x - \tfrac{1}{2}\tau) \, \varphi(x + \tfrac{1}{2}\tau) \, e^{i\theta x} \, dx, \tag{5.29}$$

$$M_2(\theta,\tau) = \Phi_2(\theta,\tau) \int \varphi^*(x - \tfrac{1}{2}\tau) \, \varphi(x + \tfrac{1}{2}\tau) \, e^{i\theta x} \, dx. \tag{5.30}$$

Divide one equation by the other to obtain

$$M_2(\theta,\tau) = \frac{\Phi_2(\theta,\tau)}{\Phi_1(\theta,\tau)} M_1(\theta,\tau). \tag{5.31}$$

5.3. Relation Between Distributions

To obtain the relation in terms of distributions, use Eq. (5.14) to obtain

$$C_2(x,p) = \frac{1}{4\pi^2} \iint \frac{\Phi_2(\theta,\tau)}{\Phi_1(\theta,\tau)} M_1(\theta,\tau) e^{-i\theta x - i\tau p} d\theta\, d\tau \qquad (5.32)$$

and further

$$C_2(x,p) = \frac{1}{4\pi^2} \iiiint \frac{\Phi_2(\theta,\tau)}{\Phi_1(\theta,\tau)} C_1(x',p') e^{i\theta(x'-x) + i\tau(p'-p)} d\theta\, d\tau\, dx'\, dp'. \qquad (5.33)$$

This relationship can be written as

$$C_2(x,p) = \iint g_{21}(x'-x, p'-p)\, C_1(x',p')\, dx'\, dp' \qquad (5.34)$$

with

$$g_{21}(x,p) = \frac{1}{4\pi^2} \iint \frac{\Phi_2(\theta,\tau)}{\Phi_1(\theta,\tau)} e^{i\theta x + i\tau p} d\theta\, d\tau. \qquad (5.35)$$

Using Eq. (2.74) or (2.75), the relation between $C_2(x,p)$ and $C_1(x,p)$ can be expressed in operational form,

$$C_2(x,p) = \frac{\Phi_2\left(i\frac{\partial}{\partial x}, i\frac{\partial}{\partial p}\right)}{\Phi_1\left(i\frac{\partial}{\partial x}, i\frac{\partial}{\partial p}\right)} C_1(x,p). \qquad (5.36)$$

If we take $C_1(x,p)$ to be the Wigner distribution, the kernel, Φ_1, is 1 and we have

$$C(x,p) = \Phi\left(i\frac{\partial}{\partial x}, i\frac{\partial}{\partial p}\right) W(x,p) \qquad (5.37)$$

where Φ is the kernel of $C(x,p)$.

Example. In the next chapter we will study a number of distributions, among them the case called standard ordering where the kernel is

$$\Phi_S(\theta,\tau) = e^{-i\theta\tau/2}. \qquad (5.38)$$

When this kernel is substituted into Eq. (5.2) one obtains

$$C_S(x,p) = \frac{1}{2\pi} \varphi^*(x) \int e^{-i\tau p} \varphi(x+\tau)\, d\tau = \frac{1}{\sqrt{2\pi}} \varphi^*(x) e^{ipx} \widehat{\varphi}(p). \qquad (5.39)$$

The Wigner distribution is obtained by taking the kernel equal to 1, $\Phi_W(\theta,\tau) = 1$,

$$W(x,p) = \frac{1}{2\pi} \int \varphi^*(x - \tfrac{1}{2}\tau) e^{-ip\tau} \varphi(x + \tfrac{1}{2}\tau)\, d\tau. \qquad (5.40)$$

To illustrate Eq. (5.36) we have that

$$C_S(x,p) = \frac{\Phi_S\left(i\frac{\partial}{\partial x}, i\frac{\partial}{\partial p}\right)}{\Phi_W\left(i\frac{\partial}{\partial x}, i\frac{\partial}{\partial p}\right)} C_W(x,p) = \exp\left(\frac{i}{2}\frac{\partial}{\partial x}\frac{\partial}{\partial p}\right) W(x,p) \qquad (5.41)$$

and inversely

$$W(x,p) = \exp\left(-\frac{i}{2}\frac{\partial}{\partial x}\frac{\partial}{\partial p}\right) C_S(x,p). \qquad (5.42)$$

5.4 Manifestly Positive Distributions

Consider the kernel

$$\Phi_{SP}(\theta,\tau) = \int h^*(x - \tfrac{1}{2}\tau)\, e^{-i\theta x}\, h(x + \tfrac{1}{2}\tau)\, dx \qquad (5.43)$$

where $h(x)$ is an arbitrary function called the window function. Upon substitution into Eq. (5.2) one obtains

$$C_{SP}(x,p) = \left|\frac{1}{\sqrt{2\pi}} \int e^{-ip\tau} \varphi(\tau)\, h(\tau - x)\, d\tau\right|^2. \qquad (5.44)$$

This distribution, called the spectrogram, does not satisfy the marginal conditions Eqs. (5.27) and (5.28). However there are other manifestly positive distributions which are not bilinear and do satisfy the marginals.

Also, one can express Eq. (5.44) in terms of the Fourier transform of the window. If we define

$$\widehat{h}(p) = \frac{1}{\sqrt{2\pi}} \int e^{-ipx} h(x)\, dx, \qquad (5.45)$$

$$h(x) = \frac{1}{\sqrt{2\pi}} \int e^{ipx} \widehat{h}(p)\, dp, \qquad (5.46)$$

then it follows that

$$\Phi_{SP}(\theta,\tau) = \int \widehat{h}^*(p - \tfrac{1}{2}\theta) e^{ip\tau} \widehat{h}(p + \tfrac{1}{2}\theta)\, dp \qquad (5.47)$$

and

$$C_{SP}(x,p) = \left|\frac{1}{\sqrt{2\pi}} \int e^{i\theta x} \widehat{\varphi}(\theta)\, \widehat{h}(p - \theta)\, d\theta\right|^2. \qquad (5.48)$$

5.5 Singular Kernels

Many of the relations in this chapter and in Chap. 4 and the specific cases we consider in Chap. 6 have kernels that are singular. Hence, many of the relations have to be interpreted and manipulated as generalized functions, that is, by using delta functions. This issue has been carefully studied by Sala, Palao, and Muga [75] who have devised methods to handle the singularities.

Chapter 6

Special Cases

In this chapter we discuss a number of associations that have been studied over the years and we show how they can be studied in a direct manner using the method we developed in previous chapters. For the sake of clarity we restate the fundamental formulas for the general association and summarize the basic results.

6.1 Summary

Generalized Correspondence Operator. For the symbol $a(x,p)$ with Fourier transform $\hat{a}(\theta, \tau)$,

$$A^\Phi(X, D) = \iint \hat{a}(\theta, \tau) \Phi(\theta, \tau) e^{i\theta X + i\tau D} \, d\theta \, d\tau \tag{6.1}$$

$$= \iint \hat{a}(\theta, \tau) \Phi(\theta, \tau) e^{i\theta\tau/2} e^{i\theta X} e^{i\tau D} \, d\theta \, d\tau \tag{6.2}$$

$$= \iint \hat{a}(\theta, \tau) \Phi(\theta, \tau) e^{-i\theta\tau/2} e^{i\tau D} e^{i\theta X} \, d\theta \, d\tau, \tag{6.3}$$

$$A^\Phi(X, D) = \frac{1}{4\pi^2} \iiiint a(q,p) \, \Phi(\theta, \tau) \, e^{-i\theta q - i\tau p + i\theta\tau/2} \, e^{i\theta X} \, e^{i\tau D} \, d\theta \, d\tau \, dq \, dp. \tag{6.4}$$

The Operation on a Function, $\varphi(x)$.

$$A^\Phi(X, D) \varphi(x) = \iint \hat{a}(\theta, \tau) \Phi(\theta, \tau) e^{i\theta\tau/2} e^{i\theta x} \varphi(x+\tau) \, d\theta \, d\tau. \tag{6.5}$$

In terms of the symbol

$$A^\Phi[\varphi] = \frac{1}{4\pi^2} \iiiint a(q,p) \, e^{-i\theta(q - x - \tau/2) - i\tau p} \, \Phi(\theta, \tau) \varphi(\tau + x) \, d\tau \, dq \, dp \, d\theta. \tag{6.6}$$

Monomial. If $a(x,p) = x^n p^m$,

$$x^n p^m \leftrightarrow \frac{1}{i^n i^m} \frac{\partial^{n+m}}{\partial \theta^n \partial \tau^m} \Phi(\theta,\tau) e^{i\theta\tau/2} e^{i\theta X} e^{i\tau D}\Big|_{\theta,\tau=0} \quad (6.7)$$

$$= \frac{1}{i^n i^m} \frac{\partial^{n+m}}{\partial \theta^n \partial \tau^m} \Phi(\theta,\tau) e^{-i\theta\tau/2} e^{i\tau D} e^{i\theta X}\Big|_{\theta,\tau=0}. \quad (6.8)$$

Rearrangement operator. $R^\Phi(x,p)$ is defined by

$$R^\Phi(x,p) = \text{rearrange } A^\Phi(X,D), \text{ so that all the } X \text{ factors are to the left of the } D \text{ operators; then replace } (X,D) \text{ by } (x,p) \quad (6.9)$$

and is explicitly given by

$$R^\Phi(x,p) = \exp\left(\frac{1}{2i}\frac{\partial}{\partial x}\frac{\partial}{\partial p}\right) \Phi\left(\frac{1}{i}\frac{\partial}{\partial x}, \frac{1}{i}\frac{\partial}{\partial p}\right) a(x,p) \quad (6.10)$$

and

$$R^\Phi(x,p) = \iint \hat{a}(\theta,\tau) \Phi(\theta,\tau) e^{i\theta\tau/2} e^{i\theta x} e^{i\tau p} \, d\theta \, d\tau. \quad (6.11)$$

From Operator to Symbol.

$$\hat{a}(\theta,\tau) = \frac{1}{4\pi^2} \frac{e^{-i\theta\tau/2}}{\Phi(\theta,\tau)} \iint R^\Phi(x,p) e^{-i\theta x - i\tau p} \, dx \, dp \quad (6.12)$$

and,

$$a(x,p) = \frac{\exp\left[-\frac{1}{2i}\frac{\partial}{\partial x}\frac{\partial}{\partial p}\right]}{\Phi\left(\frac{1}{i}\frac{\partial}{\partial x}, \frac{1}{i}\frac{\partial}{\partial p}\right)} R^\Phi(x,p). \quad (6.13)$$

Generalized distribution.

$$C(x,p) = \frac{1}{4\pi^2} \iiint \varphi^*(q - \tfrac{1}{2}\tau)\varphi(q + \tfrac{1}{2}\tau) \Phi(\theta,\tau) e^{-i\theta x - i\tau p + i\theta q} \, d\theta \, d\tau \, dq. \quad (6.14)$$

6.2 Standard Ordering

Standard ordering is where

$$x^n p^m \leftrightarrow X^n D^m \quad (6.15)$$

and

$$e^{i\theta x + i\tau p} \leftrightarrow e^{i\theta X} e^{i\tau D}. \quad (6.16)$$

6.2. Standard Ordering

This ordering corresponds to the kernel

$$\Phi_S(\theta, \tau) = e^{-i\theta\tau/2}. \tag{6.17}$$

We study this ordering in some detail as it will illustrate the various relations we have previously derived. Using Eq. (6.2) we obtain

$$A^S(X, D) = \iint \hat{a}(\theta, \tau) e^{i\theta X} e^{i\tau D} d\theta \, d\tau \tag{6.18}$$

and for the operation on a function $\varphi(x)$, we use Eq. (6.5) to obtain

$$A^S(X, D) \varphi(x) = \iint \hat{a}(\theta, \tau) e^{i\theta x} \varphi(x + \tau) \, d\theta \, d\tau. \tag{6.19}$$

In terms of the symbol we use Eq. (6.6) and after some manipulations we obtain

$$A^S(X, D) \varphi(x) = \frac{1}{2\pi} \iint a(x, p) \, e^{-i(\tau - x)p} \varphi(\tau) \, dp \, d\tau. \tag{6.20}$$

Also

$$A^S(X, D) = \frac{1}{2\pi} \iint a(X, p) \, e^{-i\tau p} e^{i\tau D} \, dp \, d\tau. \tag{6.21}$$

Consider the rearrangement operator approach as given by Eq. (6.10). We have that

$$\Phi_S\left(\frac{1}{i}\frac{\partial}{\partial x}, \frac{1}{i}\frac{\partial}{\partial p}\right) = \exp\left(-\frac{1}{2i}\frac{\partial}{\partial x}\frac{\partial}{\partial p}\right) \tag{6.22}$$

and therefore

$$R^S(x, p) = a(x, p). \tag{6.23}$$

This is a simple but powerful result. It says that to obtain the operator merely take the symbol, $a(x, p)$, rearrange it so that the x factors are to the left of the p factors and then substitute X and D. In particular, for the symbol

$$a(x, p) = f(x)h(p) \tag{6.24}$$

where $f(x)$ and $h(p)$ are any functions, one immediately obtains

$$A^S(X, D) = f(X)h(D). \tag{6.25}$$

That is, for standard ordering

$$f(x)h(p) \leftrightarrow f(X)h(D). \tag{6.26}$$

To illustrate the use of the monomial formula, Eq. (6.7), we have that

$$C_{nm}(X, D) = \frac{1}{i^n i^m} \frac{\partial^{n+m}}{\partial \theta^n \partial \tau^m} e^{i\theta X} e^{i\tau D} \bigg|_{\theta, \tau = 0} \tag{6.27}$$

$$= X^n D^m. \tag{6.28}$$

For the distribution, substitute the kernel into Eq. (6.14) to obtain

$$C_S(x,p) = \frac{1}{\sqrt{2\pi}} \varphi^*(x) e^{ipx} \widehat{\varphi}(p). \qquad (6.29)$$

This distribution was first proposed by Kirkwood [43], and studied by Margenau and Hill [51], Mehta [56], Rihaczek [72], and others.

6.3 Anti-Standard Ordering

For anti-standard ordering one takes

$$\Phi_{AS}(\theta, \tau) = e^{i\theta\tau/2}. \qquad (6.30)$$

Using Eq. (6.8)

$$A^{AS}(X, D) = \iint \widehat{a}(\theta, \tau) e^{i\tau D} e^{i\theta X} d\theta d\tau \qquad (6.31)$$

we see that

$$e^{i\theta x + i\tau p} \leftrightarrow e^{i\tau D} e^{i\theta X} \qquad (6.32)$$

and

$$x^n p^m \leftrightarrow D^m X^n. \qquad (6.33)$$

For the operation on a function, substitute Eq. (6.30) into (6.5) to obtain

$$A^{AS}(X, D) \varphi(x) = \iint \widehat{a}(\theta, \tau) e^{i\theta(\tau+x)} \varphi(\tau + x) d\theta d\tau. \qquad (6.34)$$

In terms of the symbol we use Eq. (6.6) to obtain

$$A^{AS}(X, D) \varphi(x) = \frac{1}{2\pi} \iint a(\tau, p) e^{-i(\tau-x)p} \varphi(\tau) d\tau dp \qquad (6.35)$$

and

$$A^{AS}(X, D) = \frac{1}{2\pi} \iint a(\tau + X, p) e^{-i\tau p} e^{i\tau D} d\tau dp. \qquad (6.36)$$

For the distribution one obtains

$$C_{AS}(x,p) = \frac{1}{\sqrt{2\pi}} \varphi(x) e^{-ipx} \widehat{\varphi}^*(p) \qquad (6.37)$$

which is seen to be the complex conjugate of $C_S(x,p)$,

$$C_{AS}(x,p) = C_S^*(x,p). \qquad (6.38)$$

6.4 Symmetrization Ordering

If one symmetrizes the standard and anti-standard rules by adding the kernels and dividing by two we obtain

$$\Phi_{MH}(\theta,\tau) = \cos\tfrac{1}{2}\theta\tau \tag{6.39}$$

which is often called the Margenau-Hill association because of their study of this case [51]. Using the same methods as above it is straightforward to obtain that

$$e^{i\theta x + i\tau\xi} \leftrightarrow \tfrac{1}{2}\left[e^{i\tau D}e^{i\theta X} + e^{i\theta X}e^{i\tau D}\right] \tag{6.40}$$

and

$$x^n p^m \leftrightarrow \tfrac{1}{2}\left(X^n D^m + D^m X^n\right). \tag{6.41}$$

The operator for an arbitrary symbol is

$$A^{MH}(X,D) = \tfrac{1}{2}\iint \widehat{a}(\theta,\tau)\left[e^{i\tau D}e^{i\theta X} + e^{i\theta X}e^{i\tau D}\right] d\theta\, d\tau \tag{6.42}$$

$$= \tfrac{1}{2}\iint \widehat{a}(\theta,\tau)\cos\tfrac{1}{2}\theta\tau\, e^{i\theta X + i\tau D}\, d\theta\, d\tau. \tag{6.43}$$

In terms of the symbol we have,

$$A^{MH}(X,D) = \frac{1}{4\pi}\iint [a(x,p) + a(x+\tau,p)]\, e^{-i\tau p}\, e^{i\tau D}\, dp\, d\tau \tag{6.44}$$

and for the operation on any function we obtain

$$A^{MH}[\varphi] = \frac{1}{4\pi}\iint [a(x,p) + a(\tau,p)]\, e^{-i(\tau-x)p}\, \varphi(\tau)\, dp\, d\tau. \tag{6.45}$$

For the distribution we have

$$C_{MH}(x,p) = \tfrac{1}{2}[C_S(x,p) + C_{AS}(x,p)] = \frac{1}{\sqrt{2\pi}}\operatorname{Re}\left\{\varphi^*(x)e^{ipx}\widehat{\varphi}(p)\right\}. \tag{6.46}$$

6.5 Born-Jordan Ordering

As early as 1926, Born and Jordan [10], who were two of the inventors of quantum mechanics, addressed the issue of how to write operators corresponding to a classical function. They proposed the association

$$x^n p^m \leftrightarrow \frac{1}{m+1}\sum_{\ell=0}^{m} D^{m-\ell}X^n D^{\ell} = \frac{1}{n+1}\sum_{\ell=0}^{n} X^{n-\ell}D^m X^{\ell}. \tag{6.47}$$

As we showed in Chap. 4, Eq. (4.108), this corresponds to the kernel

$$\Phi(\theta,\tau) = \operatorname{sinc}\tfrac{1}{2}\theta\tau. \tag{6.48}$$

In Section 6.8 on the ZAM kernel, we consider a somewhat more general case. Therefore, we give here the results for the Born-Jordan rule and discuss the derivations in the ZAM section. The operator and the operation on a state function are given by, respectively,

$$A^{BJ}(X,D) = \frac{1}{2\pi} \int \frac{e^{-i\tau p}}{|\tau|} \left\{ \int_{X-|\tau|/2+\tau/2}^{X+|\tau|/2+\tau/2} a(q,p)dq \right\} e^{i\tau D} d\tau dp \qquad (6.49)$$

and

$$A^{BJ}(X,D)\varphi(x) = \frac{1}{2\pi} \int \frac{e^{-i\tau p}}{|\tau|} \left\{ \int_{x-|\tau|/2+\tau/2}^{x+|\tau|/2+\tau/2} a(q,p)dq \right\} \varphi(x+\tau) d\tau dp. \qquad (6.50)$$

The distribution is

$$C_{BJ}(x,p) = \frac{1}{2\pi} \int \frac{e^{-i\tau p}}{|\tau|} \int_{x-|\tau|/2}^{x+|\tau|/2} \varphi^*(q-\tfrac{1}{2}\tau)\varphi(q+\tfrac{1}{2}\tau)dq\, d\tau. \qquad (6.51)$$

6.6 Choi-Williams Ordering

Choi and Williams and Jeong and Williams [11, 39] devised a kernel that has a number of advantages over the Wigner distribution. The kernel is

$$\Phi_{CW}(\theta,\tau) = e^{-\theta^2\tau^2/\sigma} \qquad (6.52)$$

where σ is a positive constant. Note that as $\sigma \to \infty$, the kernel approaches one, which is the kernel for the Weyl case. For the operator, Eq. (6.2), we have

$$A^{CW}(X,D) = \iint \hat{a}(\theta,\tau) e^{-\theta^2\tau^2/\sigma + i\theta\tau/2}\, e^{i\theta X}\, e^{i\tau D} d\theta\, d\tau. \qquad (6.53)$$

In terms of the symbol,

$$A^{CW}(X,D) = \frac{1}{4\pi^2} \iint a(q,p)\, e^{-i\theta q - i\tau p} e^{-\theta^2\tau^2/\sigma}\, e^{i\theta\tau/2}\, e^{i\theta X}\, e^{i\tau D} d\theta\, d\tau\, dq\, dp \qquad (6.54)$$

$$= \frac{1}{4\pi^2} \iint \sqrt{\frac{\pi}{\tau^2/\sigma}} a(q,p) \exp\left[-\frac{(q-x-\tau/2)^2}{4\tau^2/\sigma}\right] e^{-i\tau p} e^{i\tau D} d\tau\, dq\, dp \qquad (6.55)$$

and for the operation on a function, Eq. (6.14), we have

$$A^{\Phi}[\varphi] = \frac{1}{4\pi^2} \iint \sqrt{\frac{\pi}{\tau^2/\sigma}} a(q,p) \exp\left[-\frac{(q-x-\tau/2)^2}{4\tau^2/\sigma}\right] e^{-i\tau p} \varphi(\tau+x) d\tau\, dq\, dp. \qquad (6.56)$$

6.7. Weyl Ordering

For the distribution we obtain

$$C_{CW}(x,p) = \frac{1}{4\pi^{3/2}} \iint \frac{1}{\sqrt{\tau^2/\sigma}} \exp\left[-\frac{(q-x)^2}{4\tau^2/\sigma} - i\tau p\right] \varphi^*(x-\tfrac{1}{2}\tau)\, \varphi(x+\tfrac{1}{2}\tau)\, dq\, d\tau. \tag{6.57}$$

Also, using Eq. (6.10) we have

$$R^{CW}(x,p) = \exp\left[-\frac{1}{\sigma}\frac{\partial^2}{\partial x^2}\frac{\partial^2}{\partial p^2} + \frac{1}{2i}\frac{\partial}{\partial x}\frac{\partial}{\partial p}\right] a(x,p). \tag{6.58}$$

Example. Consider the case

$$a(x,p) = xp. \tag{6.59}$$

We have

$$R^{CW}(x,p)xp = \exp\left[\frac{1}{2i}\frac{\partial}{\partial x}\frac{\partial}{\partial p}\right] xp = xp - \tfrac{i}{2} \tag{6.60}$$

and hence

$$A^{CW}(X,D) = XD - \tfrac{i}{2} = \tfrac{1}{2}[XD + DX]. \tag{6.61}$$

Thus the Choi-Williams association for xp is the same as the Weyl case.

Example. Now consider the case

$$a(x,p) = x^2 p^2. \tag{6.62}$$

First, we have that

$$\exp\left[-\frac{1}{\sigma}\frac{\partial^2}{\partial x^2}\frac{\partial^2}{\partial p^2}\right] x^2 p^2 = x^2 p^2 - \tfrac{4}{\sigma} \tag{6.63}$$

giving

$$R^{CW}(x,p)x^2p^2 = \exp\left[\frac{1}{2i}\frac{\partial}{\partial x}\frac{\partial}{\partial p}\right](x^2p^2 - \tfrac{4}{\sigma}) = x^2p^2 - 2ixp - \tfrac{1}{2} - \tfrac{4}{\sigma} \tag{6.64}$$

and hence

$$A^{CW}(X,D) = X^2 D^2 - 2iXD - \tfrac{1}{2} - \tfrac{4}{\sigma}. \tag{6.65}$$

6.7 Weyl Ordering

Although we have derived the main results for the Weyl operator in Chap. 3 it is of some interest to rederive some of them as a special case of the generalized method. The Weyl case is obtained by taking

$$\Phi_W(\theta, \tau) = 1. \tag{6.66}$$

From Eq. (6.1) and (6.7) we have

$$A^W(X, D) = \iint \hat{a}(\theta, \tau) e^{i\theta\tau/2} e^{i\theta X} e^{i\tau D} d\theta \, d\tau \qquad (6.67)$$

and

$$A^W(X, D)\varphi(\tau) = \frac{1}{2\pi} \iint a\left(\frac{\tau + x}{2}, p\right) e^{i(x-\tau)p} \varphi(\tau) \, d\tau \, dp. \qquad (6.68)$$

Using Eq. (6.7) we have

$$A^W(X, D) = \frac{1}{i^n i^m} \frac{\partial^{n+m}}{\partial \theta^n \partial \tau^m} e^{i\theta\tau/2} e^{i\theta X} e^{i\tau D} \bigg|_{\theta,\tau=0} \qquad (6.69)$$

$$= \frac{1}{2^m} \sum_{\ell=0}^{m} \binom{m}{\ell} D^{m-\ell} X^n D^{\ell} \qquad (6.70)$$

$$= \frac{1}{2^n} \sum_{\ell=0}^{n} \binom{n}{\ell} X^{n-\ell} D^m X^{\ell}. \qquad (6.71)$$

The rearrangement operator, Eq. (6.10), gives

$$R_A(x, p) = \exp\left[\frac{1}{2i} \frac{\partial^2}{\partial x \partial p}\right] a(x, p) \qquad (6.72)$$

and for the distribution Eq. (6.10) we obtain

$$W(x, p) = \frac{1}{2\pi} \int \varphi^*(x - \tfrac{1}{2}\tau) e^{-ip\tau} \varphi(x + \tfrac{1}{2}\tau) \, d\tau \qquad (6.73)$$

which is the Wigner distribution.

6.8 ZAM Ordering

Zhao, Atlas, and Marks [94, 65] proposed a distribution that has many interesting properties and has found many applications. We consider first a slightly more general kernel,

$$\Phi_Z(\theta, \tau) = g(\tau) |\tau| \operatorname{sinc}(\gamma \theta \tau) \qquad (6.74)$$

where $g(\tau)$ is a function that can control the placement of the so-called cross terms in the distribution. The Born-Jordan distribution is obtained for $g(\tau) = 1/|\tau|$ and $\gamma = \tfrac{1}{2}$ and the original ZAM distribution is obtained for $g(\tau) = \tau/|\tau|$. Substituting $\Phi_Z(\theta, \tau)$ into Eq. (6.4) we have that the operator is

$$A^Z(X, D) = \frac{1}{4\pi^2} \iiiint a(q, p) g(\tau) |\tau| \operatorname{sinc}(\gamma \theta \tau) e^{-i\theta(q - X - \tau/2)} e^{i\tau D} d\theta \, d\tau \, dq \, dp. \qquad (6.75)$$

6.8. ZAM Ordering

Using

$$\int \operatorname{sinc}(\gamma\theta\tau)\, e^{-i\theta\eta}\, d\theta = \begin{cases} \dfrac{\pi}{\gamma|\tau|}, & -\gamma|\tau| \leq \eta \leq \gamma|\tau| \\ 0, & \text{otherwise} \end{cases} \quad (6.76)$$

with $\eta = q - X - \tau/2$, Eq. (6.75) simplifies to

$$A^Z(X, D) = \frac{1}{4\pi\gamma} \int g(\tau) e^{-i\tau p} \left\{ \int_{X-\gamma|\tau|+\tau/2}^{X+\gamma|\tau|+\tau/2} a(q, p)\, dq \right\} e^{i\tau D}\, d\tau\, dp. \quad (6.77)$$

The operation on a function gives

$$A^Z(X, D)\varphi(x) = \frac{1}{4\pi\gamma} \int g(\tau) e^{-i\tau p} \left\{ \int_{x-\gamma|\tau|+\tau/2}^{x+\gamma|\tau|+\tau/2} a(q, p)\, dq \right\} \varphi(x + \tau)\, d\tau\, dp. \quad (6.78)$$

Now consider the distribution, Eq. (6.14),

$$C_Z(x, p) = \frac{1}{4\pi^2} \iiint \varphi^*(q - \tfrac{1}{2}\tau) g(\tau)\, |\tau|\, \operatorname{sinc}(\gamma\theta\tau)\, e^{i\theta(q-x)} e^{-i\tau p} \varphi(q + \tfrac{1}{2}\tau)\, d\theta\, d\tau\, dq \quad (6.79)$$

and using Eq. (6.76) above with $\eta = x - q$ we obtain

$$C_Z(x, p) = \frac{1}{4\pi\gamma} \int g(\tau) e^{-i\tau p} \int_{x-\gamma|\tau|}^{x+\gamma|\tau|} \varphi^*(q - \tfrac{1}{2}\tau)\varphi(q + \tfrac{1}{2}\tau)\, dq\, d\tau. \quad (6.80)$$

Expectation value. Suppose that we want to find the expectation value of the symbol $a(x, p)$ which of course is given by

$$\langle a(x, p) \rangle = \int C(x, p) a(x, p)\, dx\, dp \quad (6.81)$$

$$= \frac{1}{4\pi\gamma} \int a(x, p) \int g(\tau) e^{-i\tau p} \int_{x-\gamma|\tau|}^{x+\gamma|\tau|} \varphi^*(q - \tfrac{1}{2}\tau)\varphi(q + \tfrac{1}{2}\tau)\, dq\, d\tau\, dx\, dp. \quad (6.82)$$

However from Eq. (6.78) we also have

$$\langle A^Z(X, D) \rangle = \int \varphi^*(x) A^Z(X, D) \varphi(x)\, dx, \quad (6.83)$$

$$\langle A^Z(X, D) \rangle$$

$$= \frac{1}{4\pi\gamma} \int \varphi^*(x) \int g(\tau) e^{-i\tau p} \left\{ \int_{x-\gamma|\tau|+\tau/2}^{x+\gamma|\tau|+\tau/2} a(q, p)\, dq \right\} \varphi(x + \tau)\, d\tau\, dp\, dx \quad (6.84)$$

$$= \frac{1}{4\pi\gamma} \iint g(\tau) e^{-i\tau p} \left[\varphi^*(x - \tfrac{1}{2}\tau) \left\{ \int_{x-\gamma|\tau|}^{x+\gamma|\tau|} a(q, p)\, dq \right\} \varphi(x + \tfrac{1}{2}\tau) \right] d\tau\, dp\, dx. \quad (6.85)$$

Eqs. (6.82) and (6.85) should be the same but it is not obvious that they are. But indeed it can be shown that

$$\int \left[\varphi^*(x - \tfrac{1}{2}\tau) \left\{ \int_{x-\gamma|\tau|}^{x+\gamma|\tau|} a(q,p)dq \right\} \varphi(x + \tfrac{1}{2}\tau) \right] dx \qquad (6.86)$$

$$= \int a(x,p) \int_{x-\gamma|\tau|}^{x+\gamma|\tau|} \varphi^*(q - \tfrac{1}{2}\tau)\varphi(q + \tfrac{1}{2}\tau) dq\, dx \qquad (6.87)$$

which does make them the same.

Born-Jordan: For the Born and Jordan case we take $g(\tau) = 1/|\tau|$ and $\gamma = \tfrac{1}{2}$ we obtain Eqs. (6.47)-(6.51).

ZAM distribution. Another particular case is to take $g(\tau) = 1$ and $\gamma = \tfrac{1}{2}$ which corresponds to the original ZAM distribution and the kernel is

$$\Phi_{\text{ZAM}}(\theta, \tau) = |\tau|\operatorname{sinc}(\gamma\theta\tau) = \frac{\sin\gamma\theta\tau}{\gamma\theta}. \qquad (6.88)$$

The operator is then

$$A^{\text{ZAM}}(X,D) = \frac{1}{4\pi^2} \int a(q,p)\,\tau\operatorname{sinc}(\gamma\theta\tau)\,e^{-i\theta(q-x-\tau/2)}\,e^{i\tau D}\,d\theta\,d\tau\,dq\,dp \qquad (6.89)$$

$$= \frac{1}{4\pi\gamma} \int e^{-i\tau p} \left\{ \int_{x-\gamma|\tau|/2+\tau/2}^{x+\gamma|\tau|/2+\tau/2} a(q,p)dq \right\} e^{i\tau D}\,d\tau\,dp \qquad (6.90)$$

and

$$A^{\text{ZAM}}(X,D)\varphi(x) = \frac{1}{4\pi\gamma} \int e^{-i\tau p} \left\{ \int_{x-\gamma|\tau|/2+\tau/2}^{x+\gamma|\tau|/2+\tau/2} a(q,p)dq \right\} \varphi(x+\tau)\,d\tau\,dp. \qquad (6.91)$$

For the distribution we obtain,

$$C_{ZAM}(x,p) = \frac{1}{2\pi} \int e^{-i\tau p} \int_{x-|\tau|/2}^{x+|\tau|/2} \varphi^*(q - \tfrac{1}{2}\tau)\,\varphi(q + \tfrac{1}{2}\tau)\,dq\,d\tau. \qquad (6.92)$$

6.9 Spectrogram

The kernel for the spectrogram type operator is

$$\Phi_{SP}(\theta,\tau) = \int h^*(x - \tfrac{1}{2}\tau)\,e^{-i\theta x}h(x + \tfrac{1}{2}\tau)\,dx \qquad (6.93)$$

6.9. Spectrogram

where $h(x)$ is the window function. We normalize the window function so that

$$\int |h(x)|^2 \, dx = 1. \tag{6.94}$$

For the operator we have

$$A^{SP}(X,D) = \iiiint \hat{a}(\theta,\tau) h^*(q - \tfrac{1}{2}\tau) h(q + \tfrac{1}{2}\tau) \, e^{-i\theta q} e^{i\theta \tau/2} \, e^{i\theta X} \, e^{i\tau D} dq \, d\theta \, d\tau. \tag{6.95}$$

In terms of the symbol we use Eq. (6.4) to obtain

$$A^{SP}(X,D) = \frac{1}{2\pi} \iiint a(q,p) h^*(x-q) h(x-q+\tau) e^{-i\tau p} \, e^{i\tau D} d\tau \, dq \, dp \tag{6.96}$$

and the operation on a function is

$$A^{SP}(X,D)\varphi(x) = \frac{1}{2\pi} \iiint a(q,p) h^*(X-q) h(X-q+\tau) e^{-i\tau p} \, \varphi(x+\tau) \, d\tau \, dq \, dp. \tag{6.97}$$

We have already considered the distribution in Sec. 5.4. It is manifestly positive and is given by Eq. (5.44),

$$C_{SP}(x,p) = \left| \frac{1}{\sqrt{2\pi}} \int e^{-ip\tau} \varphi(\tau) \, h(\tau-x) \, d\tau \right|^2. \tag{6.98}$$

Consider the association for x. Using Eq. (6.7) we obtain

$$x \leftrightarrow X - \langle x \rangle_h \tag{6.99}$$

where

$$\langle x \rangle_h = \int x \, |h(x)|^2 \, dx. \tag{6.100}$$

That is, in general, x does not go into X. That is the case since the kernel does not satisfy the marginal conditions, Eq. (5.27) and Eq. (5.28). Similarly

$$p \leftrightarrow D + \langle p \rangle_h \tag{6.101}$$

where

$$\langle p \rangle_h = \int p \, \left| \hat{h}(p) \right|^2 dp \tag{6.102}$$

and where $\hat{h}(p)$ is given by Eq. (5.45).

6.10 Gaussian Window

It is of interest to consider the special case of the spectrogram ordering where

$$h(x) = \left(\frac{\alpha}{\pi}\right)^{1/4} e^{-\alpha x^2/2} \qquad (6.103)$$

and using Eq. (6.93) we calculate the kernel to be

$$\Phi_G(\theta, \tau) = e^{-\alpha \tau^2/4 - \theta^2/4\alpha} \qquad (6.104)$$

and the distribution is then

$$C_G(x, p) = \sqrt{\frac{\alpha}{4\pi^3}} \left| \int \varphi(\tau) e^{-\alpha(\tau-x)^2/2 - ip\tau} d\tau \right|^2 \qquad (6.105)$$

$$= \sqrt{\frac{\alpha}{4\pi^3}} \left| \int \varphi(\tau + x) e^{-\alpha \tau^2/2 - ip\tau} d\tau \right|^2. \qquad (6.106)$$

For the operator we obtain

$$A^G(X, D) = \iint \hat{a}(\theta, \tau) e^{-\alpha \tau^2/4 - \theta^2/4\alpha} e^{i\theta\tau/2} e^{i\theta X} e^{i\tau D} d\theta \, d\tau. \qquad (6.107)$$

In terms of the symbol directly

$$A^G(X, D) = \left(\frac{\alpha}{\pi}\right)^{1/2} \iiiint a(q, p) e^{-\alpha(q - X - \tau/2)^2} e^{-i\tau p} e^{i\tau D} d\tau \, dq \, dp \, dq'. \qquad (6.108)$$

The rearrangement operator becomes

$$R^G(x, p)x = \exp\left(\frac{1}{2i}\frac{\partial}{\partial x}\frac{\partial}{\partial p} - \frac{1}{4\alpha}\frac{\partial^2}{\partial x^2} - \frac{\alpha}{4}\frac{\partial^2}{\partial p^2}\right). \qquad (6.109)$$

Example. For the association to x we use Eq. (6.13),

$$R^G(x, p)x = \exp\left(\frac{1}{2i}\frac{\partial}{\partial x}\frac{\partial}{\partial p} - \frac{1}{4\alpha}\frac{\partial^2}{\partial x^2} - \frac{\alpha}{4}\frac{\partial^2}{\partial p^2}\right) x = x \qquad (6.110)$$

and therefore

$$x \leftrightarrow X. \qquad (6.111)$$

Similarly

$$p \leftrightarrow D. \qquad (6.112)$$

These are consistent with Eqs. (6.99), and (6.102) since the expectation values, $\langle x \rangle_h$ and $\langle p \rangle_h$ for the Gaussian window are zero.

6.11. One Parameter Families

Example. Now consider the association for x^2,

$$R^G(x,p)\,x^2 = \exp\left(\frac{1}{2i}\frac{\partial}{\partial x}\frac{\partial}{\partial p} - \frac{1}{4\alpha}\frac{\partial^2}{\partial x^2} - \frac{\alpha}{4}\frac{\partial^2}{\partial p^2}\right)x^2 \qquad (6.113)$$

$$= \exp\left(-\frac{1}{4\alpha}\frac{\partial^2}{\partial x^2}\right)x^2 = x^2 - \frac{1}{2\alpha} \qquad (6.114)$$

and therefore

$$x^2 \leftrightarrow X^2 - \tfrac{1}{2\alpha}. \qquad (6.115)$$

For p^2,

$$R^G(x,p)p^2 = \exp\left(-\frac{\alpha}{4}\frac{\partial^2}{\partial p^2}\right)p^2 = p^2 - \tfrac{\alpha}{2} \qquad (6.116)$$

which shows that

$$p^2 \leftrightarrow D^2 - \tfrac{\alpha}{2}. \qquad (6.117)$$

Example. For xp we have

$$R^G(x,p)xp = \exp\left(\frac{1}{2i}\frac{\partial}{\partial x}\frac{\partial}{\partial p} - \frac{1}{4\alpha}\frac{\partial^2}{\partial x^2} - \frac{\alpha}{4}\frac{\partial^2}{\partial p^2}\right)xp \qquad (6.118)$$

$$= \exp\left(\frac{1}{2i}\frac{\partial}{\partial x}\frac{\partial}{\partial p}\right)xp = xp + \tfrac{1}{2i} \qquad (6.119)$$

and therefore

$$xp \leftrightarrow XD + \tfrac{1}{2i} = \tfrac{1}{2}(XD + DX). \qquad (6.120)$$

Furthermore for $x^2 p^2$ we have

$$\exp\left(\frac{1}{2i}\frac{\partial}{\partial x}\frac{\partial}{\partial p} - \frac{1}{4\alpha}\frac{\partial^2}{\partial x^2} - \frac{\alpha}{4}\frac{\partial^2}{\partial p^2}\right)x^2 p^2 = \left(x^2 p^2 - 2ixp - \tfrac{1}{2} - \tfrac{1}{2\alpha}p^2 - \tfrac{\alpha}{2}x^2\right) \qquad (6.121)$$

and therefore

$$x^2 p^2 \leftrightarrow X^2 D^2 - \tfrac{1}{2\alpha}D^2 - \tfrac{\alpha}{2}X^2 - 2iXD - \tfrac{1}{2}. \qquad (6.122)$$

6.11 One Parameter Families

As subclasses of the kernel method, one can devise parameter families. One such case is to take

$$\Phi_c(\theta,\tau) = e^{ic\theta\tau/2} \qquad (6.123)$$

as considered by Boggiatto, De Donno, and Oliaro [4]. For $c = 0, -1, 1$ one gets the Weyl, standard and anti-standard cases respectively. One can also obtain the Born-Jordan kernel by integrating over c,

$$\Phi_{BJ}(\theta,\tau) = \frac{1}{2}\int_{-1}^{1} \Phi_c(\theta,\tau)\,dc = \operatorname{sinc}\tfrac{1}{2}\theta\tau. \qquad (6.124)$$

The operator for a general symbol is given by

$$A^c(X, D) = \iint \hat{a}(\theta, \tau) e^{(c+1)\theta\tau/2} e^{i\theta X} e^{i\tau D} \, d\theta \, d\tau \qquad (6.125)$$

and in terms of the symbol directly one obtains

$$A^c(X, D) = \frac{1}{2\pi} \iint a\left(X + \tfrac{1+c}{2}\tau, p\right) e^{-i\tau p} e^{i\tau D} \, dp \, d\tau. \qquad (6.126)$$

For the association of $e^{i\theta x + i\tau p}$ we obtain

$$e^{i\theta x + i\tau p} \leftrightarrow e^{ic\theta\tau/2} e^{i\theta X + i\tau D} = e^{i(c+1)\theta\tau/2} e^{i\tau x} e^{i\tau D} = e^{i(c-1)\theta\tau/2} e^{i\tau D} e^{i\tau x} \qquad (6.127)$$

and the operation on a function gives

$$A^c[\varphi(x)] = \frac{1}{2\pi} \iint a\left(\tfrac{x+\tau}{2} + \tfrac{\tau-x}{2}c, p\right) e^{-i(\tau-x)p} \varphi(\tau + x) \, dp \, d\tau. \qquad (6.128)$$

The distribution works out to be

$$C_c(x, p) = \frac{1}{2\pi} \int \varphi^*(x - \tfrac{1}{2}(c+1)\tau) \varphi(x + \tfrac{1}{2}(1-c)\tau) e^{-i\tau p} \, d\tau. \qquad (6.129)$$

The special cases of standard, anti-standard and Wigner are now readily obtained by taking different values of c.

6.12 Normal and Anti-Normal Ordering

In normal ordering, one transforms to what is known as the annihilation and creation operator representation which are defined respectively by

$$\alpha = \frac{1}{\sqrt{2}}(X + iD); \qquad \alpha^\dagger = \frac{1}{\sqrt{2}}(X - iD). \qquad (6.130)$$

The inverse transformations are

$$X = \frac{1}{\sqrt{2}}(\alpha + \alpha^\dagger); \qquad D = \frac{1}{\sqrt{2i}}(\alpha - \alpha^\dagger). \qquad (6.131)$$

The operators satisfy

$$[\alpha, \alpha^\dagger] = 1. \qquad (6.132)$$

Note that α and $\alpha^\dagger$ are not Hermitian. This representation is very important in many areas of physics and chemistry and particularly in quantum electrodynamics.

The anti-normal ordering procedure is the following. For a symbol $a(x, p)$:

6.12. Normal and Anti-Normal Ordering

i) make the substitution

$$x = \frac{1}{\sqrt{2}}(\beta + \beta^*); \qquad p = \frac{1}{\sqrt{2}i}(\beta - \beta^*) \qquad (6.133)$$

giving

$$a(x,p) = a\left(\frac{1}{\sqrt{2}}(\beta + \beta^*), \frac{1}{\sqrt{2}i}(\beta - \beta^*)\right). \qquad (6.134)$$

Note that

$$\beta = \frac{1}{\sqrt{2}}(x + ip); \qquad \beta^* = \frac{1}{\sqrt{2}}(x - ip). \qquad (6.135)$$

In Eq. (6.134) we still have ordinary variables. Next,

ii) Rearrange $a(x,p)$ as given by Eq. (6.134) so that

$$\begin{cases} \beta^* \text{ factors are to the left of the } \beta \text{ factors} & \text{(normal ordering)}, \\ \beta^* \text{ factors are to the right of the } \beta \text{ factors} & \text{(anti-normal ordering)}. \end{cases} \qquad (6.136)$$

iii) Replace β, β^* with the operators $a, a^\dagger$ respectively.

iv) To express the resulting operator in terms of the X, D operators use Eq. (6.130).

We now obtain the kernel corresponding to the anti-normal case. Consider the association for $e^{i\theta x + i\tau p}$. We follow the above steps to obtain

$$e^{i\theta x + i\tau p} = e^{i\theta(\beta+\beta^*)/\sqrt{2} + i\tau(\beta-\beta^*)i/\sqrt{2}} = e^{\beta(i\theta+\tau)/\sqrt{2}} e^{\beta^*(i\theta-\tau)/\sqrt{2}}. \qquad (6.137)$$

Now all the β^* factors are to the right of the β factors. Therefore

$$e^{i\theta x + i\tau p} \leftrightarrow e^{a(i\theta+\tau)/\sqrt{2}} e^{a^\dagger(i\theta-\tau)/\sqrt{2}} \qquad (6.138)$$

and in terms of X, D we have

$$e^{i\theta x + i\tau p} \leftrightarrow e^{(X+iD)(i\theta+\tau)/2} e^{(X-iD)(i\theta-\tau)/2}. \qquad (6.139)$$

Using Eq. (6.139), simplification of the right hand side gives

$$e^{(X+iD)(i\theta+\tau)/2} e^{(X-iD)(i\theta-\tau)/2} = e^{-\frac{1}{4}[\theta^2+\tau^2]} e^{i\theta X + i\tau D} \qquad (6.140)$$

and finally

$$e^{i\theta x + i\tau p} \leftrightarrow e^{-\frac{1}{4}[\theta^2+\tau^2]} e^{i\theta X + i\tau D}. \qquad (6.141)$$

Therefore the kernel is

$$\Phi_{AN}(\theta, \tau) = e^{-(\theta^2+\tau^2)/4}. \qquad (6.142)$$

We see that this is a special case of the spectrogram with a Gaussian window, Eq. (6.103), and with $\alpha = 1$. A similar calculation leads to

$$\Phi_N(\theta, \tau) = e^{(\theta^2+\tau^2)/4}. \qquad (6.143)$$

The important properties of these correspondences is that for symbols of the form $a_1(x - ip)a_2(x + ip)$ the correspondences are

$$a_1(x - ip)a_2(x + ip) \leftrightarrow a_1(X - iD)a_2(X + iD) \qquad \text{(normal ordering)} \qquad (6.144)$$

and

$$a_1(x - ip)a_2(x + ip) \leftrightarrow a_2(X + iD)a_1(X - iD) \qquad \text{(anti-normal ordering)}. \qquad (6.145)$$

Chapter 7

Unitary Transformation

Our aim in this chapter is to study how the various results we have derived change under a unitary transformation. Of particular importance is that the generalized operator and distributions keep their functional form. Unitary transformations are important for many reasons but certainly among the most important is that many equations of evolution for a field can often be expressed as a unitary transformation on the initial field. By a field we mean a function of two variables; for example the temperature, electric field, pressure field, are functions of both space and time. To illustrate with a specific example, suppose we consider the following equation of evolution

$$i\frac{\partial}{\partial t}\varphi(x,t) = -\frac{\partial^2}{\partial x^2}\varphi(x,t) + x\varphi(x,t) \tag{7.1}$$

where we have to find $\varphi(x,t)$ given $\varphi(x,0)$. The solution can be expressed as

$$\varphi(x,t) = e^{-it\left(-\frac{\partial^2}{\partial x^2}+x\right)}\varphi(x,0). \tag{7.2}$$

That is, the operator, $e^{-it\left(-\frac{\partial^2}{\partial x^2}+x\right)}$, operating on the initial field gives the field at a later time. More generally, if we have an equation of evolution of the form

$$i\frac{\partial}{\partial t}\varphi(x,t) = H\varphi(x,t) \tag{7.3}$$

where H is a time independent Hermitian operator, then the solution is

$$\varphi(x,t) = e^{-iHt}\varphi(x,0) \tag{7.4}$$

as can be readily verified by differentiating and seeing that it leads to Eq. (7.3). The basic property of the operator, e^{-iHt}, is that it is unitary, a concept that we now explain.

A unitary operator, U, is an operator that satisfies

$$U^\dagger U = I; \qquad U^\dagger = U^{-1} \tag{7.5}$$

where $U^\dagger$ and U^{-1} are the adjoint and inverse operators respectively. There are two types of associated transformations, the transformation of an operator and the transformation of functions.

The unitary transformation of an operator from A to A' is defined by

$$A' = U^\dagger A U. \qquad (7.6)$$

This transformation is often called a similarity transformation.

Unitary transformation of the state function from φ to φ' is defined by

$$\varphi' = U^\dagger \varphi = U^{-1}\varphi. \qquad (7.7)$$

A basic property of a unitary transformation is that the eigenvalues of A' and A are the same. To see this consider

$$Au(a,x) = au(a,x). \qquad (7.8)$$

Multiply from the left by $U^\dagger$ and let $u(a,x) = Uu'(a,x)$,

$$U^\dagger A U u'(a,x) = a U^\dagger U u'(a,x). \qquad (7.9)$$

Since $U^\dagger U = 1$ we have

$$A' u'(a,x) = au'(a,x) \qquad (7.10)$$

which shows that the eigenvalues of A and A' are the same and that the eigenfunctions of A' are u' where

$$u'(a,x) = U^{-1}u(a,x) = U^\dagger u(a,x). \qquad (7.11)$$

A unitary transformation preserves the inner product of two functions. That is, for two functions φ_1 and φ_2,

$$\int \varphi_1^* \varphi_2 \, dx = \int \varphi_1^{*\prime} \varphi_2' \, dx. \qquad (7.12)$$

For the case where $\varphi_1 = \varphi_2$ a unitary transformation preserves the norm

$$\int |\varphi|^2 \, dx = \int |\varphi'|^2 \, dx. \qquad (7.13)$$

7.1 Unitary and Hermitian Operators

A unitary operator can always be written in the form

$$U = e^{-iH} \qquad (7.14)$$

where H is a Hermitian operator. Conversely, if H is Hermitian, then U is unitary. The minus sign in the exponent is inserted for convenience and for historical reasons. Therefore we can write a unitary transformation of an operator and state function respectively as

$$A' = e^{iH} A e^{-iH}, \tag{7.15}$$

$$\varphi' = e^{iH} \varphi. \tag{7.16}$$

A fundamental formula for simplifying and evaluating Eq. (7.15) is that for any two operators A and H and real number ξ,

$$e^{\xi H} A e^{-\xi H} = A + \xi[H, A] + \frac{1}{2!}\xi^2[H, [H, A]] + \frac{1}{3!}\xi^3[H, [H, [H, A]]] + \cdots. \tag{7.17}$$

7.2 Unitary Transformation of the Generalized Association

We now consider the transformation of the generalized association under a unitary transformation on X, D [6, 7]. Take

$$X' = U^\dagger X U; \qquad D' = U^\dagger D U, \tag{7.18}$$

and suppose the corresponding ordinary variables are (x', p'),

$$X' \leftrightarrow x', \tag{7.19}$$

$$D' \leftrightarrow p'. \tag{7.20}$$

Inversely, we have

$$X = U X' U^\dagger, \qquad D = U D' U^\dagger. \tag{7.21}$$

Define the generalized operator correspondence for X', D' by

$$A^\Phi(X', D') = \iint \hat{a}(\theta, \tau) \Phi(\theta, \tau) e^{i\theta X' + i\tau D'} d\theta d\tau \tag{7.22}$$

$$= \iint \hat{a}(\theta, \tau) \Phi(\theta, \tau) e^{i\theta U^\dagger X U + i\tau U^\dagger D U} d\theta d\tau \tag{7.23}$$

and consider

$$e^{i\theta U^\dagger X U + i\tau U^\dagger D U} = e^{U^\dagger (i\theta X + i\tau D) U} \tag{7.24}$$

$$= \sum_{n=0}^{\infty} \frac{i^n}{n!} \left(U^\dagger (i\theta X + i\tau D) U\right)^n. \tag{7.25}$$

Using the fact that $U^\dagger = U^{-1}$ we have that

$$\left(U^\dagger (\theta X + \tau D) U\right)^n = U^\dagger (\theta X + \tau D)^n U \tag{7.26}$$

and therefore
$$e^{i\theta U^\dagger X U + i\tau U^\dagger D U} = U^\dagger e^{i\theta X + i\tau D} U. \tag{7.27}$$

Substituting this result into Eq. (7.23) we obtain
$$A^\Phi(X', D') = U^\dagger A^\Phi(X, D) U \tag{7.28}$$

which shows that if X and D transform by a unitary transformation, then so does the general correspondence. In Eq. (7.28) one understands that a substitution using Eq. (7.21) is made on the right hand side.

7.3 Transformation of the Generalized Distribution

In a subsequent chapter we discuss the issue of obtaining joint distributions for arbitrary operators. Here we just consider the case where the two operators X' and D' are unitary transformations of X and D as given by Eq. (7.18). One can express the distribution of (x', p') in terms of the distribution of (x, p) as we now show.

The characteristic function for X' and D' for a state $\varphi(x)$ is

$$M_{x'p'}(\theta, \tau) = \Phi(\theta, \tau) \int \varphi^*(x) e^{i\theta X' + i\tau D'} \varphi(x) \, dx \tag{7.29}$$

and using Eq. (7.27) we have

$$M_{ab}(\theta, \tau) = \Phi(\theta, \tau) \int \varphi^*(x) U^\dagger e^{i\theta X + i\tau D} U \varphi(x) \, dx \tag{7.30}$$

$$= \Phi(\theta, \tau) \int (U\varphi(x))^* e^{i\theta X + i\tau D} U \varphi(x) \, dx \tag{7.31}$$

$$= \Phi(\theta, \tau) \int \varphi'^*(x) e^{i\theta X + i\tau D} \varphi'(x) \, dx. \tag{7.32}$$

Comparing this with Eq. (5.22) we see that it is exactly the characteristic function of x, p except that we are in a state $\varphi'(x)$ given by Eq. (7.16). Therefore we can immediately write

$$C(x', p') = \frac{1}{4\pi^2} \iiint \varphi'^*(q - \tfrac{1}{2}\tau) \varphi'(q + \tfrac{1}{2}\tau) \Phi(\theta, \tau) e^{-i\theta x' - i\tau p' + i\theta q} \, dq \, d\tau \, d\theta. \tag{7.33}$$

This is an important result because it allows one to immediately write distributions for new variables if their corresponding operators are connected to X and D by a unitary transformation. All one has to do is change the state function by way of Eq. (7.7).

7.4 Examples

Example. Translation operator. Take

$$U = e^{i\ell D} \tag{7.34}$$

where ℓ is a real number. We have that

$$U^\dagger = U^{-1} = e^{-i\ell D} \tag{7.35}$$

and the transformation of a function is

$$\varphi'(x) = e^{-i\ell D}\varphi(x) = \varphi(x - \ell). \tag{7.36}$$

The operators X' and D' are

$$X' = U^\dagger X U = e^{-i\ell D} X e^{i\ell D} = X - \ell, \tag{7.37}$$

$$D' = U^\dagger D U = D. \tag{7.38}$$

For the generalized correspondence operator we have

$$A'^\Phi(X', D') = e^{-i\ell D} A^\Phi(X, D) e^{i\ell D}. \tag{7.39}$$

The distribution of x' and p' is then, according to Eq. (7.33)

$$C(x', p') = \frac{1}{4\pi^2} \iiint \varphi^*(q - \tfrac{1}{2}\tau - \ell)\, \varphi(q + \tfrac{1}{2}\tau - \ell)\, \Phi(\theta, \tau)\, e^{-i\theta x' - i\tau p' + i\theta q} dq\, d\tau\, d\theta. \tag{7.40}$$

Example. $U = e^{-itD^2}$. We take

$$U = e^{-itD^2}. \tag{7.41}$$

For the X' operator we have

$$X' = e^{itD^2} X e^{-itD^2} = X + it[-2Di] = X + 2tD \tag{7.42}$$

where we have used Eq. (7.17) and the fact that

$$[X, D^2] = 2iD. \tag{7.43}$$

For the D' operator it is clear that

$$D' = e^{itD^2} D e^{-itD^2} = D. \tag{7.44}$$

The transformation of a function is given by

$$\varphi'(x) = e^{-itD^2}\varphi(x) = \frac{1}{\sqrt{4\pi it}} \int \varphi(x') \exp\left[-\frac{(x'-x)^2}{4it}\right] dx' \tag{7.45}$$

which can be proven in the following way. Write

$$\widehat{\varphi}(p) = \frac{1}{\sqrt{2\pi}} \int \varphi(x) e^{-ixp} \, dx, \qquad (7.46)$$

$$\varphi(x) = \frac{1}{\sqrt{2\pi}} \int \widehat{\varphi}(p) e^{ixp} \, dp. \qquad (7.47)$$

Now

$$e^{-itD^2} \varphi(x) = \frac{1}{\sqrt{2\pi}} \int \widehat{\varphi}(p) e^{-itD^2} e^{ixp} \, dp \qquad (7.48)$$

$$= \frac{1}{\sqrt{2\pi}} \int \widehat{\varphi}(p) e^{-itp^2} e^{ixp} \, dp \qquad (7.49)$$

where in going from Eq. (7.48) to Eq. (7.49) we have used $e^{-itD^2} e^{ixp} = e^{-itp^2} e^{ixp}$. In Eq. (7.49) substitute for $\widehat{\varphi}(p)$ to obtain

$$e^{-itD^2} \varphi(x) = \frac{1}{2\pi} \iint \varphi(x') e^{-itp^2} e^{i(x-x')p} \, dp \, dx' \qquad (7.50)$$

which gives Eq. (7.45). The distribution is hence

$$C(x'p') = \frac{1}{4\pi^2} \iiint \Phi(\theta, \tau) \, \varphi'^*(q - \tfrac{1}{2}\tau) \, \varphi'(q + \tfrac{1}{2}\tau) \, e^{-i\theta x' - i\tau p' + i\theta q} \, dq \, d\tau \, d\theta. \qquad (7.51)$$

Chapter 8

Path Integral Approach

The path integral approach for obtaining correspondence rules arose out of the work of Feynman and Kac where they showed that the Green's function for Schrödinger's equation can be written as the sum of a functional over all paths from one time to another. It is beyond our scope to discuss the path integrals but we give here the end result as to the viewpoint of the formulation as it applies to correspondence rules [15, 27, 54, 40, 57, 85]. There are two separate approaches, corresponding to configuration space and phase-space.

8.1 Configuration Space

The basic idea is that we can express the operation on a function as

$$A^\Phi(X, D)\varphi(x) = \int k(x, x')\varphi(x')\,dx' \tag{8.1}$$

where $k(x, x')$ can be calculated with the classical symbol by summing over all paths. If the paths are parameterized by a Fourier integral, it turns out that one can write $k(x, x')$ as

$$k(x, x') = \frac{1}{2\pi}\int \bar{a}(x, x', p)\,e^{-i(x'-x)p}\,dp \tag{8.2}$$

where $\bar{a}(x, x', p)$ is an averaging function taken over the path. By taking different discretizations to $\bar{a}(x, x', p)$ one obtains different correspondence rules. For example $\bar{a}(x, x', p)$ can be approximated by $\frac{1}{2}[a(x, p) + a(x', p)]$ or $a\left(\frac{x+x'}{2}, p\right)$, among others. The only constraint is that

$$\lim_{x' \to x} \bar{a}(x, x', p) = a(x, p). \tag{8.3}$$

We now express this formulation in terms of the methods developed in Chap. 4 where we showed that

$$A^{\Phi}(X,D)\varphi(x) = \frac{1}{4\pi^2} \iiiint a\left(q + \frac{x+x'}{2}, p\right) e^{-i\theta q - i(x'-x)p}$$
$$\times \Phi(\theta, x' - x)\varphi(x')\, dx'\, dq\, dp\, d\theta. \tag{8.4}$$

Comparing with Eq. (8.2) we see that

$$k(x,x') = \frac{1}{4\pi^2} \iiint a\left(q + \frac{x+x'}{2}, p\right) e^{-i\theta q - i(x'-x)p} \Phi(\theta, x' - x)\, dq\, dp\, d\theta \tag{8.5}$$

and further

$$\bar{a}(x, x', p) = \frac{1}{2\pi} \iint a\left(q + \frac{x+x'}{2}, p\right) e^{-i\theta q} \Phi(\theta, x' - x)\, dq\, d\theta. \tag{8.6}$$

Example. Weyl rule. The kernel is 1 and we have that

$$\bar{a}(x, x', p) = \frac{1}{2\pi} \iint a\left(q + \frac{x+x'}{2}, p\right) e^{-i\theta q}\, dq\, d\theta \tag{8.7}$$

which gives

$$\bar{a}(x, x', p) = a\left(\frac{x+x'}{2}, p\right). \tag{8.8}$$

Also,

$$k(x, x') = \frac{1}{2\pi} \int a\left(\frac{x+x'}{2}, p\right) e^{i(x-x')p}\, dp. \tag{8.9}$$

Example. Symmetrization rule. The kernel is

$$\Phi(\theta, \tau) = \tfrac{1}{2} \cos \tfrac{1}{2}\theta\tau \tag{8.10}$$

and substituting this kernel into Eq. (8.5) we obtain that

$$k(x, x') = \frac{1}{4\pi} \int e^{-i(x'-x)p} \left[a(x, p) + a(x', p)\right] dp. \tag{8.11}$$

Also,

$$\bar{a}(x, x', p) = \tfrac{1}{2}\left[a(x, p) + a(x', p)\right]. \tag{8.12}$$

8.2 Phase-Space

One can also define the operator by

$$A^\Phi(X, D) = \int a(x', p') \Delta(X - x', D - p') \, dx' \, dp' \qquad (8.13)$$

where Δ can be calculated over phase-space paths. Comparing with Eq. (4.97) we have

$$\Delta(X - x', D - p') = \frac{1}{4\pi^2} \iint \Phi(\theta, \tau) e^{i\theta(X-x') + i\tau(D-p')} \, d\theta \, d\tau \qquad (8.14)$$

$$= \frac{1}{4\pi^2} \iint \Phi(\theta, \tau) e^{i\theta\tau/2} e^{i\theta(X-x')} e^{i\tau(D-p')} \, d\theta \, d\tau. \qquad (8.15)$$

Since all the X factors are to the left of the D factors we can define the rearrangement function Δ_R by

$$\Delta_R(x - x', p - p') = \frac{1}{4\pi^2} \iint e^{i\theta\tau/2} e^{i\theta(x-x')} e^{i\tau(p-p')} \Phi(\theta, \tau) \, d\theta \, d\tau \qquad (8.16)$$

or

$$\Delta_R(x, p) = \frac{1}{4\pi^2} \iint e^{i\theta(x+\tau/2)} e^{i\tau p} \Phi(\theta, \tau) \, d\theta \, d\tau. \qquad (8.17)$$

Eq. (8.17) can be inverted to express the kernel in terms of Δ_R,

$$\Phi(\theta, \tau) = e^{-i\theta\tau/2} \iint \Delta_R(x, p) e^{-i\theta x} e^{-i\tau p} \, dx \, dp. \qquad (8.18)$$

Substituting into Eq. (8.13) we obtain that

$$A^\Phi(X, D) = \frac{1}{2\pi} \iiiint a(x', p') \Delta_R(x - x', p - p') e^{-i\tau p} e^{i\tau D} \, d\tau \, dx' \, dp \, dp' \qquad (8.19)$$

and

$$A^\Phi[\varphi] = \frac{1}{2\pi} \iiiint a(x', p') \Delta_R(x - x', p - p') e^{-i\tau p} \varphi(\tau + x) \, d\tau \, dx' \, dp \, dp'. \qquad (8.20)$$

Example. Weyl rule. The kernel is 1 and we have

$$\Delta_R(x, p) = \frac{1}{4\pi^2} \iint e^{i\theta(x+\tau/2)} e^{i\tau p} \, d\theta \, d\tau \qquad (8.21)$$

$$= \frac{1}{2\pi} \int \delta(\tau/2 + x) e^{i\tau p} \, d\tau \qquad (8.22)$$

$$= \frac{1}{\pi} e^{-2ixp}. \qquad (8.23)$$

Using Eq. (8.20) we have

$$A^W[\varphi] = \frac{1}{2\pi}\frac{1}{\pi}\iiiint a(x',p')\,e^{-2i(x-x')(p-p')}e^{-i\tau p}\,\varphi(\tau+x)\,d\tau\,dx'\,dp\,dp' \quad (8.24)$$

which evaluates to

$$A^W[\varphi] = \frac{1}{2\pi}\iint a\left(\tfrac{1}{2}(x+\tau),p\right)e^{-i(\tau-x)p}\,\varphi(\tau)\,d\tau\,dp \quad (8.25)$$

which is the result for the Weyl case given in Eq. (3.16).

Example. Standard ordering. For standard ordering the kernel is $\Phi_S(\theta,\tau) = e^{-i\theta\tau/2}$ and therefore

$$\Delta_R(x,p) = \frac{1}{4\pi^2}\iint e^{i\theta x}e^{i\tau p}\,d\theta\,d\tau = \delta(x)\delta(p) \quad (8.26)$$

and substituting into Eq. (8.20) we have

$$A^\Phi[\varphi] = \frac{1}{2\pi}\iiiint a(x',p')\,\delta(x-x')\delta(p-p')e^{-i\tau p}\,\varphi(\tau+x)\,d\tau\,dx'\,dp\,dp' \quad (8.27)$$

$$= \frac{1}{2\pi}\iint a(x,p)\,e^{-i\tau p}\,\varphi(\tau+x)\,d\tau\,dp \quad (8.28)$$

which is the same as Eq. (6.20).

Chapter 9

Time-Frequency Operators

The fundamental idea of time-frequency analysis is to understand and describe how the frequency content of a signal is changing in time [20, 18]. By a signal one means a function of time, for example the electric and magnetic fields, pressure, voltage etc. Time functions are called wave forms or signals. We shall denote the signal by $s(t)$. Examples of man-made signals whose frequencies are clearly changing in time are music and human speech where indeed it is the changing frequencies that are the essence of the signals. The symbol, $a(t, \omega)$, now is a function of time and frequency. From a mathematical point of view, the methodology that has been developed in the previous chapters can be totally carried over to time-frequency.

We define the time and frequency operators by $\mathcal{T}$ and $\mathcal{W}$ where,

$$\mathcal{T} = \begin{cases} t & \text{in the time representation} \\ i\frac{d}{d\omega} & \text{in the frequency (Fourier) representation,} \end{cases} \qquad (9.1)$$

$$\mathcal{W} = \begin{cases} \frac{1}{i}\frac{d}{dt} & \text{in the } t \text{ representation} \\ \omega & \text{in the Fourier representation.} \end{cases} \qquad (9.2)$$

The fundamental relation between $\mathcal{T}$ and $\mathcal{W}$ is the commutator,

$$[\mathcal{T}, \mathcal{W}] = \mathcal{T}\mathcal{W} - \mathcal{W}\mathcal{T} = i. \qquad (9.3)$$

The spectrum of a signal, $\hat{s}(\omega)$, is defined by

$$\hat{s}(\omega) = \frac{1}{\sqrt{2\pi}} \int s(t)\, e^{-it\omega}\, dt \qquad (9.4)$$

with

$$s(t) = \frac{1}{\sqrt{2\pi}} \int \hat{s}(\omega)\, e^{it\omega}\, d\omega. \qquad (9.5)$$

The Fourier transform of the symbol is defined by

$$\hat{a}(\theta, \tau) = \frac{1}{4\pi^2} \iint a(t, \omega)\, e^{-i\theta t - i\tau\omega}\, dt\, d\omega \qquad (9.6)$$

and the symbol, $a(t,\omega)$, is given by

$$a(t,\omega) = \iint \hat{a}(\theta,\tau) e^{i\theta t + i\tau\omega} \, d\theta \, d\tau. \tag{9.7}$$

Thus, the correspondence from the notation of the previous chapters to that of time-frequency is

$$x, p \to t, \omega \tag{9.8}$$
$$X, D \to \mathcal{T}, \mathcal{W} \tag{9.9}$$
$$\varphi(x), \hat{\varphi}(p) \to s(t), \hat{s}(\omega) \tag{9.10}$$
$$a(x,p), A^\Phi(X,D) \to a(t,\omega), A^\Phi(\mathcal{T},\mathcal{W}). \tag{9.11}$$

Furthermore we point out that the (x,p) case that we have considered thus far can be called space- (spatial) frequency analysis. However, there is one fundamental difference in that signals in nature are real, and hence one has to consider how to write a complex signal corresponding to a real signal. This is discussed in Sec. 9.3

9.1 Time-Frequency Association Rules

In total analogy with the development of Chap. 4 we define the time-frequency symbol to be any function of time and frequency $a(t,\omega)$ and the corresponding operator associated with that symbol by

$$A^\Phi(\mathcal{T},\mathcal{W}) = \iint \hat{a}(\theta,\tau) \Phi(\theta,\tau) e^{i\theta\mathcal{T}+i\tau\mathcal{W}} \, d\theta \, d\tau. \tag{9.12}$$

Since

$$e^{i\theta\mathcal{T}+i\tau\mathcal{W}} = e^{i\theta\tau/2} e^{i\theta\mathcal{T}} e^{i\tau\mathcal{W}} = e^{-i\theta\tau/2} e^{i\tau\mathcal{W}} e^{i\theta\mathcal{T}} \tag{9.13}$$

we have

$$A^\Phi(\mathcal{T},\mathcal{W}) = \iint \hat{a}(\theta,\tau) \Phi(\theta,\tau) e^{i\theta\tau/2} e^{i\theta\mathcal{T}} e^{i\tau\mathcal{W}} \, d\theta \, d\tau \tag{9.14}$$

$$= \iint \hat{a}(\theta,\tau) \Phi(\theta,\tau) e^{-i\theta\tau/2} e^{i\tau\mathcal{W}} e^{i\theta\mathcal{T}} \, d\theta \, d\tau. \tag{9.15}$$

In terms of the symbol,

$$A^\Phi(\mathcal{T},\mathcal{W}) = \frac{1}{4\pi^2} \iiiint a(t,\omega) \Phi(\theta,\tau) e^{-i\theta t - i\tau\omega + i\theta\tau/2} e^{i\theta\mathcal{T}} e^{i\tau\mathcal{W}} \, d\theta \, d\tau \, dt \, d\omega. \tag{9.16}$$

The operation on a signal is

$$A^\Phi(\mathcal{T},\mathcal{W}) s(t) = \iint \hat{a}(\theta,\tau) \Phi(\theta,\tau) e^{i\theta\tau/2} e^{i\theta t} s(t+\tau) \, d\theta \, d\tau \tag{9.17}$$

9.1. Time-Frequency Association Rules

and in terms of the symbol we have

$$A^\Phi(\mathcal{T},\mathcal{W})\,s(t) = \frac{1}{4\pi^2} \iiiint a\left(q + \frac{t+\tau}{2}, \omega\right) e^{-i\theta q + i(t-\tau)\omega} \\ \times \Phi(\theta, \tau - t)\, s(\tau)\, d\tau\, dq\, d\omega\, d\theta \qquad (9.18)$$

and also

$$A^\Phi(\mathcal{T},\mathcal{W})\,s(t) \\ = \frac{1}{4\pi^2} \iiiint a(t',\omega)\, e^{-i\theta(t'-t-\tau/2) - i\tau\omega}\, \Phi(\theta,\tau)\, s(\tau+t)\, d\tau\, dt'\, d\omega\, d\theta. \qquad (9.19)$$

For the operation on a function in the Fourier domain one obtains

$$A^\Phi(\mathcal{T},\mathcal{W})\,\widehat{s}(\omega) = \iint \widehat{a}(\theta,\tau)\Phi(\theta,\tau) e^{-i\theta\tau/2}\, e^{i\tau\omega}\, \widehat{s}(\omega-\theta)\, d\theta\, d\tau \qquad (9.20)$$

and in terms of the symbol one has

$$A^\Phi(\mathcal{T},\mathcal{W})\,\widehat{s}(\omega) = \frac{1}{4\pi^2} \iiiint a(t, \omega' + \tfrac{1}{2}(\omega+\theta))\Phi(\omega-\theta,\tau) \\ \times e^{-i\tau\omega'} e^{i(\theta-\omega)t}\, \widehat{s}(\theta)\, d\theta\, d\tau\, dt\, d\omega'. \qquad (9.21)$$

For the rearrangement operator one defines

$$R^\Phi(t,\omega) = \text{rearrange } A^\Phi(\mathcal{T},\mathcal{W}), \text{ so that all the } \mathcal{T} \text{ factors are to the left} \\ \text{of the } \mathcal{W} \text{ operators; then replace } (\mathcal{T},\mathcal{W}) \text{ by } (t,\omega) \qquad (9.22)$$

giving

$$R^\Phi(t,\omega) = \iint \widehat{a}(\theta,\tau)\Phi(\theta,\tau)\, e^{i\theta\tau/2} e^{i\theta t}\, e^{i\tau\omega}\, d\theta\, d\tau \qquad (9.23)$$

which can be written in the following form,

$$R^\Phi(t,\omega) = \exp\left(\frac{1}{2i}\frac{\partial}{\partial t}\frac{\partial}{\partial \omega}\right) \Phi\left(\frac{1}{i}\frac{\partial}{\partial t}, \frac{1}{i}\frac{\partial}{\partial \omega}\right) a(t,\omega). \qquad (9.24)$$

For two different associations characterized by kernels $\Phi_1(\theta,\tau)$ and $\Phi_2(\theta,\tau)$ corresponding to $R^{\Phi_1}(t,\omega)$ and $R^{\Phi_2}(t,\omega)$ we have

$$R^{\Phi_2}(t,\omega) = \frac{\Phi_2\left(\frac{1}{i}\frac{\partial}{\partial t}, \frac{1}{i}\frac{\partial}{\partial \omega}\right)}{\Phi_1\left(\frac{1}{i}\frac{\partial}{\partial t}, \frac{1}{i}\frac{\partial}{\partial \omega}\right)} R^{\Phi_1}(t,\omega). \qquad (9.25)$$

9.2 Time-Frequency Distributions

The developments in Chapter 5 can now be re-written for the time-frequency case. One can define an infinite number of time-frequency distributions, $C(t,\omega)$, so that for a symbol $a(t,\omega)$ and a signal $s(t)$,

$$\int s^*(t) A^\Phi (\mathcal{T}, \mathcal{W}) s(t) dt = \iint a(t,\omega) C(t,\omega) dt d\omega \qquad (9.26)$$

where

$$C(t,\omega) = \frac{1}{4\pi^2} \iiint s^*(q - \tfrac{1}{2}\tau) s(q + \tfrac{1}{2}\tau) \Phi(\theta,\tau) e^{-i\theta t - i\tau\omega + i\theta q} d\theta d\tau dq. \qquad (9.27)$$

$C(t,\omega)$ is called the generalized time-frequency distribution. In terms of the Fourier transform of the signal it is given by

$$C(t,\omega) = \frac{1}{4\pi^2} \iiint \widehat{s}^*(k + \tfrac{1}{2}\theta) \widehat{s}(k - \tfrac{1}{2}\theta) \Phi(\theta,\tau) e^{-i\theta t - i\tau\omega + i\tau k} d\theta d\tau dk. \qquad (9.28)$$

Also, the characteristic function is

$$M(\theta,\tau) = \langle \Phi(\theta,\tau) e^{i\theta \mathcal{T} + i\tau \mathcal{W}} \rangle = \Phi(\theta,\tau) \int s^*(t) e^{i\theta \mathcal{T} + i\tau \mathcal{W}} s(t) dt \qquad (9.29)$$

which evaluates to

$$M(\theta,\tau) = \Phi(\theta,\tau) \int s^*(t - \tfrac{1}{2}\tau) e^{i\theta t} s(t + \tfrac{1}{2}\tau) dt. \qquad (9.30)$$

9.3 Complex Signals and Instantaneous Frequency

In our previous development, the state function could be any complex function and indeed in quantum mechanics the state function is inherently complex. However in signal analysis, signals are inherently real and dealing with real signals leads to difficulties. For example, for real signals the energy density spectrum, which is the absolute square of the spectrum, is symmetric and hence the average frequency is zero which is unsatisfactory. Therefore one seeks to define a complex signal that corresponds in some sense to the real signal at hand. In addition, if we did find a way to define the complex signal, then one can define instantaneous frequency by the derivative of the phase.

The general approach is to seek a complex signal of the form

$$z(t) = s_r(t) + i s_i(t) \qquad (9.31)$$

where one insists that the real part, s_r, is the real signal at hand

$$s_r(t) = s(t). \qquad (9.32)$$

9.3. Complex Signals and Instantaneous Frequency

Obviously there is an infinite number of ways to write $z(t)$, since the imaginary part, $s_i(t)$, is thus far arbitrary. Gabor gave a procedure to fix $s_i(t)$, and the resulting complex signal is called the analytic signal. We point out that other definitions have been given. A detailed discussion of these issues can be found in references [25, 20, 50, 68].

The Gabor procedure [32, 20] is as follows. For a real signal, $s(t)$, the spectrum, $\hat{s}(\omega)$, has both positive and negative frequencies; define the complex signal, $z(t)$, whose spectrum is composed of the positive frequencies of $\hat{s}(\omega)$ only. Hence

$$z(t) = 2 \frac{1}{\sqrt{2\pi}} \int_0^\infty \hat{s}(\omega) e^{i\omega t} dt \qquad (9.33)$$

where the factor of 2 is inserted so that the real part of $z(t)$ will be $s(t)$. In Eq. (9.33) substitute for $\hat{s}(\omega)$ to obtain

$$z(t) = 2 \frac{1}{2\pi} \int \int_0^\infty s(t') e^{i(t-t')\omega} d\omega dt'. \qquad (9.34)$$

Using a basic property of the delta function,

$$\int_0^\infty e^{i(t-t')\omega} d\omega = \pi \, \delta(t - t') + \frac{i}{t - t'} \qquad (9.35)$$

we have

$$z(t) = \frac{1}{\pi} \int s(t') \left[\pi \, \delta(t - t') + \frac{i}{t - t'} \right] dt' \qquad (9.36)$$

giving

$$z(t) = s(t) + \frac{i}{\pi} \int \frac{s(t')}{t - t'} dt'. \qquad (9.37)$$

The second part of Eq. (9.37) is the Hilbert transform of $s(t)$ and one writes

$$H[s(t)] = \frac{1}{\pi} \int \frac{s(t')}{t - t'} dt'. \qquad (9.38)$$

The integration in Eq. (9.38) implies taking the principle part. Therefore, we can write the analytic signal as

$$z(t) = s(t) + iH[s(t)] = z_r(t) + i z_i(t) \qquad (9.39)$$

where

$$z_r(t) = s(t); \qquad z_i(t) = H[s(t)]. \qquad (9.40)$$

Having defined the complex signal one writes

$$z(t) = A(t) e^{i\varphi(t)} \qquad (9.41)$$

and defines the amplitude and phase of a signal by

$$A(t) = \sqrt{z_r^2 + z_i^2}; \qquad \varphi(t) = \arctan z_i/z_r. \qquad (9.42)$$

This allows one to then define the instantaneous frequency by

$$\omega_i(t) = \frac{d}{dt}\varphi(t). \qquad (9.43)$$

9.4 Time-Frequency Space-(Spatial) Frequency

Having defined time-frequency distributions and space-(spatial) frequency distributions, it is natural to consider symbols and distributions of the four variables $a(x, p, t, \omega)$ and state functions that are fields, that is functions of space and time, $\varphi(x, t)$. We give just an outline how this can be carried out for the Weyl case. We define the Fourier transform of the state function by

$$\widehat{\varphi}(p, \omega) = \frac{1}{2\pi} \int \varphi(x, t) \, e^{-ixp - it\omega} \, dt \, dx, \qquad (9.44)$$

$$\varphi(x, t) = \frac{1}{2\pi} \int \widehat{\varphi}(p, \omega) \, e^{ixp + it\omega} d\omega \, dp, \qquad (9.45)$$

and for the four-dimensional symbol, $a(x, p, t, \omega)$, we define

$$\widehat{a}(\theta_x, \tau_x, \theta, \tau) = \left(\frac{1}{4\pi^2}\right)^2 \iiiint a(x, p, t, \omega) e^{-i\theta_x x - i\tau_x p} e^{-i\theta t - i\tau \omega} \, dx \, dp \, dt \, d\omega, \qquad (9.46)$$

$$a(x, p, t, \omega) = \iiiint \widehat{a}(\theta_x, \tau_x, \theta, \tau) e^{i\theta_x x + i\tau_x p} e^{i\theta t + i\tau \omega} \, d\theta_x \, d\tau_x \, d\theta \, d\tau. \qquad (9.47)$$

The four-dimensional operator is then

$$A(X, D, T, W) = \iiiint \widehat{a}(\theta_x, \tau_x, \theta, \tau) \, e^{i\theta_x X + i\tau_x D} \, e^{i\theta T + i\tau W} d\theta_x \, d\tau_x \, d\theta \, d\tau \qquad (9.48)$$

and the operation on a function is given by

$$A(X, D, T, W)\varphi(x,t) = \iiiint \widehat{a}(\theta_x, \tau_x, \theta, \tau) \, e^{i\theta_x \tau_x/2 + i\theta \tau/2} \, e^{i\theta_x x + i\theta t}$$
$$\times \varphi(x + \tau_x, t + \tau) \, d\theta_x \, d\tau_x \, d\theta \, d\tau. \qquad (9.49)$$

The four-dimensional Wigner distribution is

$$W(x, p, t, \omega) = \left(\frac{1}{2\pi}\right)^2 \iint \varphi^*(x - \tfrac{1}{2}\tau_x, t - \tfrac{1}{2}\tau) \, \varphi(x + \tfrac{1}{2}\tau_x, t + \tfrac{1}{2}\tau) \, e^{-i\tau_x p - i\tau \omega} \, d\tau \, d\tau_x \qquad (9.50)$$

9.4. Time-Frequency Space-(Spatial) Frequency

and this definition assures that

$$\iint \varphi^*(x,t) A(X, D, \mathcal{T}, \mathcal{W}) \, \varphi(x,t) dx dt$$
$$= \iiiint a(x, p, t, \omega) W(x, p, t, \omega) \, dx \, dp \, dt \, d\omega. \tag{9.51}$$

Furthermore with these definitions the appropriate marginal conditions are satisfied in that

$$\int W(x, p, t, \omega) \, d\omega = W(x, p; t), \tag{9.52}$$

$$\int W(x, p, t, \omega) \, dp = W(t, \omega; x), \tag{9.53}$$

where $W(x,p)$ and $W(t,\omega)$ are Wigner distributions in the variables indicated and in this case defined by

$$W(x, p; t) = \frac{1}{2\pi} \int \varphi^*(x - \tfrac{1}{2}\tau_x, t) \, \varphi(x + \tfrac{1}{2}\tau_x, t) \, e^{-i\tau_x p} \, d\tau_x, \tag{9.54}$$

$$W(t, \omega; x) = \frac{1}{2\pi} \int \varphi^*(x, t - \tfrac{1}{2}\tau) \, \varphi(x, t + \tfrac{1}{2}\tau) \, e^{-i\tau\omega} \, d\tau. \tag{9.55}$$

Chapter 10

Transformation of Differential Equations Into Phase Space

Suppose we want to obtain the Wigner distribution of a function $\varphi(x)$ which is the solution of a linear differential equation

$$a_n \frac{d^n \varphi(x)}{dx^n} + a_{n-1} \frac{d^{n-1} \varphi(x)}{dx^{n-1}} \cdots + a_1 \frac{d\varphi(x)}{dx} + a_0 \varphi(x) = f(x). \qquad (10.1)$$

One could solve for $\varphi(x)$ and then substitute into the Wigner distribution

$$W(x,p) = \frac{1}{2\pi} \int \varphi^*(x - \tfrac{1}{2}\tau) e^{-i\tau p} \varphi(x + \tfrac{1}{2}\tau) \, d\tau. \qquad (10.2)$$

In this chapter we show that one can bypass solving the differential equation for $\varphi(x)$ and obtain the corresponding differential equation for the Wigner distribution directly [29, 30]. The main reason one wants to do that is that in phase-space, the nature of the solution is often revealed much more clearly than in x or p space.

10.1 Transformation Properties of the Wigner Distribution

Suppose we have a Wigner distribution of $\varphi(x)$. Then, what is the Wigner distribution of $\varphi'(x)$ or the Wigner distribution of $g(x)\varphi(x)$, where $g(x)$ is some function? In this section we derive such properties as they are important to obtain the differential equation for the Wigner distribution. It is convenient to define the cross Wigner distribution of two functions φ and ψ by

$$W_{\varphi,\psi} = \frac{1}{2\pi} \int \varphi^*(x - \tfrac{1}{2}\tau) e^{-i\tau p} \psi(x + \tfrac{1}{2}\tau) \, d\tau. \qquad (10.3)$$

Chapter 10. Transformation of Differential Equations Into Phase Space

For the sake of neatness we use the following notation for differentiation:

$$\varphi'(x) = \frac{d}{dx}\varphi(x), \qquad \varphi^{(n)} = \frac{d^n}{dx^n}\varphi(x). \tag{10.4}$$

Consider $W_{\varphi,\psi'}$,

$$W_{\varphi,\psi'} = \frac{1}{2\pi}\int \varphi^*(x-\tfrac{1}{2}\tau)e^{-i\tau p}\frac{\partial}{\partial x}\psi(x+\tfrac{1}{2}\tau)\,d\tau \tag{10.5}$$

$$= \frac{2}{2\pi}\int \varphi^*(x-\tfrac{1}{2}\tau)e^{-i\tau p}\frac{\partial}{\partial \tau}\psi(x+\tfrac{1}{2}\tau)\,d\tau \tag{10.6}$$

$$= -\frac{2}{2\pi}\int \frac{\partial}{\partial \tau}\left\{\varphi^*(x-\tfrac{1}{2}\tau)e^{-i\tau p}\right\}\psi(x+\tfrac{1}{2}\tau)\,d\tau \tag{10.7}$$

$$= \frac{1}{2\pi}\int \left[\frac{\partial}{\partial x}\varphi^*(x-\tfrac{1}{2}\tau) + 2ip\varphi^*(x-\tfrac{1}{2}\tau)\right]e^{-i\tau p}\psi(x+\tfrac{1}{2}\tau)\,d\tau \tag{10.8}$$

and therefore

$$W_{\varphi,\psi'} = W_{\varphi',\psi} + 2ipW_{\varphi,\psi}. \tag{10.9}$$

Also, it is clear that

$$\frac{\partial}{\partial x}W_{\varphi,\psi} = W_{\varphi,\psi'} + W_{\varphi',\psi}. \tag{10.10}$$

From these two equations we obtain

$$W_{\varphi',\psi} = \left(\frac{1}{2}\frac{\partial}{\partial x} - ip\right)W_{\varphi,\psi} = A_x W_{\varphi,\psi} \tag{10.11}$$

and

$$W_{\varphi,\psi'} = \left(\frac{1}{2}\frac{\partial}{\partial x} + ip\right)W_{\varphi,\psi} = B_x W_{\varphi,\psi} \tag{10.12}$$

where we have defined

$$A_x = \frac{1}{2}\frac{\partial}{\partial x} - ip, \qquad B_x = \frac{1}{2}\frac{\partial}{\partial x} + ip. \tag{10.13}$$

Combining Eq. (10.11) and Eq. (10.12) we have

$$W_{\varphi^{(n)},\psi^{(m)}} = A^n B^m W_{\varphi,\psi} \tag{10.14}$$

and

$$W_{\varphi^{(n)},\varphi^{(n)}} = A^n B^n W_{\varphi,\varphi} = \left(\frac{1}{4}\frac{\partial^2}{\partial x^2} + p^2\right)^n W_{\varphi,\varphi}. \tag{10.15}$$

Suppose we have the cross Wigner distribution of $\varphi(x)$ and $\psi(x)$ and wish to obtain the cross Wigner distributions of $g(x)\varphi(x)$ and $\psi(x)$, or $\varphi(x)$ and $g(x)\psi(x)$, where $g(x)$ is an arbitrary function. Using the above methods results in

$$W_{g\varphi,\psi}(x,p) = g^*(E_x)W_{\varphi,\psi}, \tag{10.16}$$

$$W_{\varphi,g\psi}(x,p) = g(F_x)W_{\varphi,\psi}, \qquad (10.17)$$

where

$$E_x = x + \frac{1}{2i}\frac{\partial}{\partial p}, \qquad F_x = x - \frac{1}{2i}\frac{\partial}{\partial p}. \qquad (10.18)$$

10.2 Ordinary Differential Equations

We consider ordinary differential equations of nth order, with constant coefficients

$$a_n \frac{d^n \varphi(x)}{dx^n} + a_{n-1}\frac{d^{n-1}\varphi(x)}{dx^{n-1}} \cdots + a_1 \frac{d\varphi(x)}{dx} + a_0\varphi(x) = f(x). \qquad (10.19)$$

We write the equation in polynomial notation

$$P(\mathcal{D})\varphi(x) = f(x) \qquad (10.20)$$

where

$$\mathcal{D} = \frac{d}{dx} \qquad (10.21)$$

and

$$P(\mathcal{D}) = a_n \mathcal{D}^n + a_{n-1}\mathcal{D}^{n-1} + \cdots + a_1\mathcal{D} + a_0. \qquad (10.22)$$

Taking the Wigner distribution of both sides of Eq. (10.20) we obtain

$$W_{P(\mathcal{D})\varphi, P(\mathcal{D})\varphi}(x, p) = W_{f,f}(x, p) \qquad (10.23)$$

where $W_{f,f}(x,p)$ is the Wigner distribution of the forcing term

$$W_{f,f}(x, p) = \frac{1}{2\pi}\int f^*(x - \tfrac{1}{2}\tau)e^{-i\tau p}f(x + \tfrac{1}{2}\tau)\, d\tau. \qquad (10.24)$$

Using the results described in the previous section one obtains the differential equation for the Wigner distribution

$$P^*(A)P(B)W_{\varphi,\varphi}(x, p) = W_{f,f}(x, p) \qquad (10.25)$$

where A and B are given by Eq. (10.13). Explicitly

$$P^*\left(\frac{1}{2}\frac{\partial}{\partial x} - ip\right) P\left(\frac{1}{2}\frac{\partial}{\partial x} + ip\right) W_{\varphi,\varphi}(x, p) = W_{f,f}(x, p). \qquad (10.26)$$

This equation is in general a partial differential equation of twice the order of the original differential equation, that is order $2n$. In Eq. (10.25),

$$P(A) = a_n A^n + a_{n-1}A^{n-1} + \cdots + a_1 A + a_0 \qquad (10.27)$$

and

$$P^*(A) = a_n^* A^n + a_{n-1}^* A^{n-1} + \cdots + a_1^* A + a_0^*. \qquad (10.28)$$

That is, in $P^*(A)$ only the coefficients are complex conjugated.

If the driving force is zero then
$$P^*(A)P(B)W_{\varphi,\varphi}(x,p) = 0 \qquad (10.29)$$
For this case the equation simplifies to two equations
$$P^*(A)W_{\varphi,\varphi}(x,p) = P(B)W_{\varphi,\varphi}(x,p) = 0. \qquad (10.30)$$

Example. Consider the equation
$$\frac{d^2\varphi(x)}{dx^2} + 2\mu\frac{d\varphi(x)}{dx} + \omega_0^2\varphi(x) = f(x). \qquad (10.31)$$
Using the above procedure one obtains
$$\left[a_4\frac{\partial^4}{\partial x^4} + a_3\frac{\partial^3}{\partial x^3} + a_2\frac{\partial^2}{\partial x^2} + a_1\frac{\partial}{\partial x} + a_0\right]W_{\varphi,\varphi}(x,p) = W_{f,f}(x,p) \qquad (10.32)$$
where,
$$a_0 = (\omega_0^2 - \omega^2)^2 + 4\mu^2\omega^2, \qquad (10.33)$$
$$a_1 = 2\mu(\omega_0^2 + \omega^2), \qquad a_2 = \frac{1}{2}(\omega_0^2 + \omega^2 + 2\mu^2), \qquad (10.34)$$
$$a_3 = \frac{1}{2}\mu, \qquad a_4 = 1/16. \qquad (10.35)$$
While Eq. (10.32) is clearly more complicated than Eq. (10.31) it turns out that the solution to Eq. (10.32) is simple and very insightful [29, 30].

10.3 Non-Constant Coefficients

For an ordinary differential equation with non-constant coefficients,
$$a_n(x)\frac{d^n\varphi(x)}{dx^n} + a_{n-1}(x)\frac{d^{n-1}\varphi(x)}{dx^{n-1}} + \cdots + a_1(x)\frac{d\varphi(x)}{dx} + a_0(x)\varphi(x) = f(x) \quad (10.36)$$
we again rewrite it in polynomial notation
$$P(\mathcal{D},x)\varphi(x) = f(x) \qquad (10.37)$$
where now
$$P(\mathcal{D},x) = a_n(x)\mathcal{D}^n + a_{n-1}(x)\mathcal{D}^{n-1} + \ldots + a_1(x)\mathcal{D} + a_0(x). \qquad (10.38)$$
A similar derivation that led to Eq. (10.25) leads now to
$$P^*(A_x, E_x)P(B_x, F_x)W_{u,u}(t,\omega) = W_{f,f}(t,\omega). \qquad (10.39)$$

Partial differential Equations. One can also transform partial differential equations into phase space equations but we do not do so here [29]. We point out that indeed in Wigner's original paper, he transformed the Schrödinger equation into a phase space equation for the Wigner distribution.

Chapter 11

The Eigenvalue Problem in Phase-Space

Suppose we have a correspondence between the symbol $a(x,p)$ and the Weyl operator $A(X,D)$,
$$a(x,p) \leftrightarrow A(X,D). \tag{11.1}$$

If $a(x,p)$ is real, then $A(X,D)$ is Hermitian and we can then consider the eigenvalue problem
$$A(X,D)u_n(x) = \lambda_n u_n(x) \tag{11.2}$$

where λ_n and $u_n(x)$ are the eigenvalues and eigenfunctions respectively. Since the operator is Hermitian, the eigenvalues are real and the eigenfunctions are complete and orthogonal. In Eq. (3.16) we showed that for an arbitrary function

$$A(X,D)u_n(x) = \iint \widehat{a}(\theta,\tau)\, e^{i\theta\tau/2}\, e^{i\theta x} u_n(x+\tau)\, d\theta\, d\tau \tag{11.3}$$

$$= \frac{1}{2\pi} \iint a\left(\frac{x+\tau}{2},p\right) e^{-ip\tau}\, u_n(\tau+x)\, d\tau\, dp. \tag{11.4}$$

Hence, the eigenvalue problem becomes

$$\frac{1}{2\pi} \iint a\left(\frac{x+\tau}{2},p\right) e^{-ip\tau}\, u_n(\tau+x)\, d\tau\, dp = \lambda_n u_n(x) \tag{11.5}$$

or

$$\iint \widehat{a}(\theta,\tau)\, e^{i\theta\tau/2}\, e^{i\theta x} u_n(x+\tau)\, d\theta\, d\tau = \lambda_n u_n(x). \tag{11.6}$$

This can be thought as an integral equation for $u_n(x)$.

We now formulate the eigenvalue problem in phase space. In Eq (11.6) let $x \to x + \tau'/2$, then multiply by $\frac{1}{2\pi} u_k^*(x - \tau'/2) e^{-i\tau' p}$, and integrate with respect to $d\tau'$ to obtain

$$\frac{1}{2\pi}\iiint u_k^*(x-\tau'/2)e^{-i\tau'p}\hat{a}(\theta,\tau)\,e^{i\theta\tau/2}\,e^{i\theta(x+\tau'/2)}u_n(x+\tau'/2+\tau)\,d\theta\,d\tau\,d\tau'$$

$$= \lambda_n \frac{1}{2\pi}\int u_k^*(x-\tau'/2)u_n(x+\tau'/2)e^{-i\tau'p}\,d\tau' \qquad (11.7)$$

The right hand side is the cross Wigner distribution

$$W_{kn}(x,p) = \frac{1}{2\pi}\int u_k^*(x-\tau/2)u_n(x+\tau/2)e^{-i\tau p}\,d\tau \qquad (11.8)$$

and therefore

$$\frac{1}{2\pi}\iiint \hat{a}(\theta,\tau)u_k^*(x-\tau'/2)\,u_n(x+\tau'/2+\tau)$$

$$\times e^{-i\tau'p}e^{i\theta\tau/2}\,e^{i\theta(x+\tau'/2)}d\theta\,d\tau\,d\tau' = \lambda_n W_{kn}(x,p) \qquad (11.9)$$

which after a few straightforward transformations gives

$$\iint W_{kn}(x+\tau/2, p-\theta/2)\hat{a}(\theta,\tau)\,e^{i\theta x+i\tau p}\,d\theta\,d\tau = \lambda_n W_{kn}(x,p). \qquad (11.10)$$

Recalling the properties of the translation operators we have that

$$\iint W_{kn}(x+\tau/2, p-\theta/2)e^{i\theta x+i\tau p}d\theta\,d\tau = e^{i\theta\left(x+\frac{i}{2}\frac{\partial}{\partial p}\right)+i\tau\left(p-\frac{i}{2}\frac{\partial}{\partial x}\right)}W_{kn}(x,p) \qquad (11.11)$$

and therefore

$$\iint \hat{a}(\theta,\tau)\,e^{i\theta\left(x+\frac{i}{2}\frac{\partial}{\partial p}\right)+i\tau\left(p-\frac{i}{2}\frac{\partial}{\partial x}\right)}W_{kn}(x,p)d\theta\,d\tau = \lambda_n W_{kn}(x,p) \qquad (11.12)$$

giving

$$a\left(x+\frac{i}{2}\frac{\partial}{\partial p}, p-\frac{i}{2}\frac{\partial}{\partial x}\right)W_{kn}(x,p) = \lambda_n W_{kn}(x,p). \qquad (11.13)$$

Example. Consider the case

$$a(x,p) = p \qquad (11.14)$$

then Eq. (11.13) becomes

$$pW_{\lambda\lambda'}(x,p) + \frac{1}{2i}\frac{\partial}{\partial x}W_{\lambda\lambda'}(x,p) = \lambda W_{\lambda\lambda'} \qquad (11.15)$$

where we have substituted λ, λ' for k, n since the eigenvalues are clearly continuous. The solution is

$$W_{\lambda\lambda'}(x,p) = \frac{1}{2\pi}e^{i(\lambda-\lambda')x}\delta(p-\lambda'/2-\lambda/2). \qquad (11.16)$$

This can be verified directly by finding the eigenfunctions, namely

$$\varphi_\lambda(x) = \frac{1}{\sqrt{2\pi}}e^{i\lambda x} \qquad (11.17)$$

and substituting into Eq. (11.8)

11.1 General Kernel

To obtain the eigenvalue equations for a general kernel we start with the generalized correspondence rule, Eq. (4.15),

$$A(X,D)\,u_n(x) = \iint \hat{a}(\theta,\tau)\,\Phi(\theta,\tau)e^{i\theta\tau/2}\,e^{i\theta x}\,u_n(x+\tau)\,d\theta\,d\tau \tag{11.18}$$

and therefore the eigenvalue problem becomes

$$\iint \hat{a}(\theta,\tau)\,\Phi(\theta,\tau)e^{i\theta\tau/2}\,e^{i\theta x}u_n(x+\tau)\,d\theta\,d\tau = \lambda_n u_n(x). \tag{11.19}$$

In Eq. (11.19) let $x \to q + \tau'/2$, then multiply by

$$\left(\frac{1}{2\pi}\right)^2 u_k^*(q-\tau'/2)\,e^{-i\theta'x - i\tau'p + i\theta'q}\,\Phi(\theta',\tau') \tag{11.20}$$

and integrate with respect $d\theta'\,d\tau'\,dq$ to obtain that

$$\left(\frac{1}{2\pi}\right)^2 \iint \hat{a}(\theta,\tau)\,u_k^*(q-\tau'/2)\,e^{-i\theta'x-i\tau'p+i\theta'q}\,\Phi(\theta',\tau')\Phi(\theta,\tau)$$
$$\times\,e^{i\theta\tau/2}\,e^{i\theta(q+\tau'/2)}u_n(q+\tau'/2+\tau)\,d\theta\,d\tau\,d\theta'\,d\tau'\,dq = \lambda_n C_{kn}(x,p) \tag{11.21}$$

where

$$C_{kn}(x,p) = \frac{1}{4\pi^2}\iiint \Phi(\theta,\tau)u_k^*(q-\tfrac{1}{2}\tau)\,u_n(q+\tfrac{1}{2}\tau)e^{-i\theta x - i\tau p + i\theta q}\,d\theta\,d\tau\,dq. \tag{11.22}$$

Furthermore using

$$u_k^*(q-\tfrac{1}{2}\tau)u_n(q+\tfrac{1}{2}\tau) = \frac{1}{2\pi}\iiint \frac{C_{kn}(x',p')e^{i\theta x' + i\tau p'}}{\Phi(\theta,\tau)}e^{-i\theta q}\,d\theta\,dx'\,dp' \tag{11.23}$$

or in general

$$u_k^*(q')u_n(q) = \frac{1}{2\pi}\iiint \frac{C_{kn}(x',p')}{\Phi(\theta,q-q')}e^{i\theta(x'-(q'+q)/2)+i(q-q')p'}\,d\theta\,dx'\,dp' \tag{11.24}$$

a straightforward but somewhat lengthy manipulation of Eq. (11.21) leads to

$$\left(\frac{1}{2\pi}\right)^2 \iint \hat{a}(\theta,\tau)\,e^{i\theta'(x'-x-\tau/2)}\,e^{i\tau'(p'-p+\theta/2)}C_{kn}(x',p')e^{i\theta x'+i\tau p'}$$
$$\times\,\frac{\Phi(\theta',\tau')\Phi(\theta,\tau)}{\Phi(\theta+\theta',\tau'+\tau)}\,d\theta\,d\tau\,d\theta'\,d\tau'\,dq\,dx'\,dp' = \lambda_n C_{kn}(x,p). \tag{11.25}$$

Using the same type of steps that led to Eq. (11.13) leads to

$$\left(\frac{1}{2\pi}\right)^2 \iint \frac{\Phi(\frac{1}{i}\frac{\partial}{\partial x'},\frac{1}{i}\frac{\partial}{\partial p'})\Phi(\frac{1}{i}\frac{\partial}{\partial x'_a},\frac{1}{i}\frac{\partial}{\partial p'_a})}{\Phi(\frac{1}{i}\frac{\partial}{\partial x'_a}+\frac{1}{i}\frac{\partial}{\partial x'},\frac{1}{i}\frac{\partial}{\partial p'_a}+\frac{1}{i}\frac{\partial}{\partial p'})} e^{i\theta'(x'-x)} e^{i\tau'(p'-p)} \widehat{a}(\theta,\tau)$$
$$\times e^{i\theta\left(x'+\frac{i}{2}\frac{\partial}{\partial p'}\right)+i\tau\left(p'-\frac{i}{2}\frac{\partial}{\partial x'}\right)} C_{kn}(x',p') \, d\theta \, d\tau \, d\theta' \, d\tau' \, dx' \, dp' = \lambda_n C_{kn}(x,p) \tag{11.26}$$

which gives

$$\frac{\Phi(\frac{1}{i}\frac{\partial}{\partial x},\frac{1}{i}\frac{\partial}{\partial p})\Phi(\frac{1}{i}\frac{\partial}{\partial x_a},\frac{1}{i}\frac{\partial}{\partial p_a})}{\Phi(\frac{1}{i}\frac{\partial}{\partial x_a}+\frac{1}{i}\frac{\partial}{\partial x},\frac{1}{i}\frac{\partial}{\partial p_a}+\frac{1}{i}\frac{\partial}{\partial p})} a\left(x+\frac{i}{2}\frac{\partial}{\partial p},p-\frac{i}{2}\frac{\partial}{\partial x}\right) C_{kn}(x,p) = \lambda_n C_{kn}(x,p). \tag{11.27}$$

If one takes $\Phi = 1$ then the Wigner case is obtained and these equations reduce to the ones of the previous section.

Example. For the kernel

$$\Phi^{AS}(\theta,\tau) = e^{i\theta\tau/2} \tag{11.28}$$

which is the anti-standard ordering rule, we have that

$$\frac{\Phi(\theta',\tau')\Phi(\theta,\tau)}{\Phi(\theta+\theta',\tau'+\tau)} = e^{-i\theta'\tau/2} e^{-i\theta\tau'/2} \tag{11.29}$$

and hence

$$\frac{\Phi(\frac{1}{i}\frac{\partial}{\partial x},\frac{1}{i}\frac{\partial}{\partial p})\Phi(\frac{1}{i}\frac{\partial}{\partial x_a},\frac{1}{i}\frac{\partial}{\partial p_a})}{\Phi(\frac{1}{i}\frac{\partial}{\partial x_a}+\frac{1}{i}\frac{\partial}{\partial x},\frac{1}{i}\frac{\partial}{\partial p_a}+\frac{1}{i}\frac{\partial}{\partial p})} = \exp\left[\frac{i}{2}\left(\frac{\partial}{\partial x}\frac{\partial}{\partial p_a}+\frac{\partial}{\partial x_a}\frac{\partial}{\partial p}\right)\right]. \tag{11.30}$$

Eq. (11.27) then becomes

$$\exp\left[\frac{i}{2}\left(\frac{\partial}{\partial x}\frac{\partial}{\partial p_a}+\frac{\partial}{\partial x_a}\frac{\partial}{\partial p}\right)\right] a\left(x+\frac{i}{2}\frac{\partial}{\partial p},p-\frac{i}{2}\frac{\partial}{\partial x}\right) C_{kn}^{AS}(x,p) = \lambda_n C_{kn}^{AS}(x,p) \tag{11.31}$$

and for Eq. (11.26) we have

$$\left(\frac{1}{2\pi}\right)^2 \iint \widehat{a}(\theta,\tau) \, e^{i\theta'x'} e^{i\tau'p'} e^{i\theta x'+i\tau p'}$$
$$\times C_{kn}^{AS}(x'+x+\tau,p'+p) \, d\theta \, d\tau \, d\theta' \, d\tau' \, dx' \, dp' = \lambda_n C_{kn}^{AS}(x,p). \tag{11.32}$$

Chapter 12

Arbitrary Operators: Single Operator

In this chapter we study the operator

$$\mathcal{M}(\theta) = e^{i\theta A} \tag{12.1}$$

where A is an arbitrary Hermitian operator and θ is real. We use calligraphic $\mathcal{M}(\theta)$ for the operator as we will use $M(\theta)$ for the characteristic function which is an ordinary function. We call $\mathcal{M}(\theta)$ the characteristic function operator because its expectation value will be a proper characteristic function in the sense that it yields a proper probability distribution. By a proper probability distribution, we mean that it is manifestly positive and normalized to 1. Also, instead of probability one can think of intensity and the same considerations apply. The basic idea is that the expectation value of $\mathcal{M}(\theta)$ in a state $\varphi(x)$,

$$M(\theta) = \int \varphi^*(x) \mathcal{M}(\theta) \varphi(x) \, dx \tag{12.2}$$

will define a proper characteristic function and associated probability distribution.

12.1 The Probability Distribution Corresponding to an Operator

Suppose we have a probability distribution, $P(\lambda)$, of the random variable λ; then the characteristic function is defined by

$$M(\theta) = \int e^{i\theta\lambda} P(\lambda) d\lambda = \langle e^{i\theta\lambda} \rangle \tag{12.3}$$

and conversely knowing $M(\theta)$ we obtain the probability distribution by way of Fourier inversion

$$P(\lambda) = \frac{1}{2\pi} \int M(\theta)\, e^{-i\theta\lambda}\, d\theta \qquad (12.4)$$

The moments are given by

$$\langle \lambda^n \rangle = \frac{1}{i^n} \frac{d^n}{d\theta^n} M(\theta) \big|_{\theta=0}. \qquad (12.5)$$

If the moments are known, the characteristic function can be constructed by way of

$$M(\theta) = \langle e^{i\theta\lambda} \rangle = \sum_{n=0}^{\infty} \frac{(i\theta)^n}{n!} \langle \lambda^n \rangle. \qquad (12.6)$$

The characteristic function is an average, the average of $e^{i\theta\lambda}$, and hence for a state $\varphi(x)$ we take

$$M(\theta) = \int \varphi^*(x)\, e^{i\theta A}\, \varphi(x)\, dx. \qquad (12.7)$$

The state function is normalized to 1,

$$\int |\varphi(x)|^2 dx = 1. \qquad (12.8)$$

Applying Eq. (12.4) the probability distribution is

$$P(\lambda) = \frac{1}{2\pi} \int M(\theta)\, e^{-i\theta\lambda} d\theta \qquad (12.9)$$

$$= \frac{1}{2\pi} \iint \varphi^*(x) e^{i\theta A} \varphi(x) e^{-i\theta\lambda}\, dx\, d\theta. \qquad (12.10)$$

There are two cases to consider, corresponding to when the spectrum of A is continuous or discrete.

Continuous case. Since A is Hermitian, the eigenvalues are real and the eigenfunctions are complete. We write

$$Au(\lambda, x) = \lambda u(\lambda, x) \qquad (12.11)$$

and expand the state function as

$$\varphi(x) = \int F(\lambda) u(\lambda, x)\, d\lambda \qquad (12.12)$$

where

$$F(\lambda) = \int \varphi(x) u^*(\lambda, x)\, d\lambda. \qquad (12.13)$$

12.1. The Probability Distribution Corresponding to an Operator

The characteristic function, Eq. (12.7), is then

$$M(\theta) = \int F^*(\lambda) u^*(\lambda, x) e^{i\theta A} F(\lambda') u(\lambda', x) \, d\lambda \, d\lambda' \, dx \tag{12.14}$$

$$= \int F^*(\lambda) u^*(\lambda, x) e^{i\theta \lambda'} F(\lambda') u(\lambda', x) \, d\lambda \, d\lambda' \, dx \tag{12.15}$$

$$= \int F^*(\lambda) e^{i\theta \lambda'} F(\lambda') \delta(\lambda - \lambda') \, d\lambda \, d\lambda' \tag{12.16}$$

which evaluates to

$$M(\theta) = \int |F(\lambda)|^2 e^{i\theta \lambda} \, d\lambda. \tag{12.17}$$

Therefore, the probability distribution is clearly

$$P(\lambda) = |F(\lambda)|^2 = \left| \int \varphi(x) u^*(\lambda, x) \, dx \right|^2 \tag{12.18}$$

which is manifestly positive.

Discrete case. For the discrete case we write

$$A u_n(x) = \lambda_n u_n(x) \tag{12.19}$$

and expand the state function, $\varphi(x)$, as

$$\varphi(x) = \sum_{n=0}^{\infty} c_n u_n(x) \tag{12.20}$$

where

$$c_n = \int u_n^*(x) \varphi(x) \, dx. \tag{12.21}$$

Substitution into Eq. (12.10) gives

$$M(\theta) = \sum_n |c_n|^2 e^{i\theta \lambda_n}. \tag{12.22}$$

Further, using Eq. (12.4) we have

$$P(\lambda) = \frac{1}{2\pi} \int M(\theta) e^{-i\theta \lambda} d\theta \tag{12.23}$$

$$= \frac{1}{2\pi} \int \sum_n |c_n|^2 e^{i\theta \lambda_n} e^{-i\theta \lambda} d\theta \tag{12.24}$$

giving

$$P(\lambda) = \sum_n |c_n|^2 \delta(\lambda_n - \lambda). \tag{12.25}$$

Therefore, the only values which are not zero are the λ_n's and the probability for obtaining λ_n is

$$P(\lambda_n) = |c_n|^2 = \left| \int u_n^*(x) \varphi(x) \, dx \right|^2. \tag{12.26}$$

12.2 Expectation Value

We now calculate the expectation value of λ. By the usual definition, it is given by

$$\langle \lambda \rangle = \int \lambda P(\lambda) d\lambda = \int \lambda |F(\lambda)|^2 d\lambda. \qquad (12.27)$$

However, one can also obtain it from the characteristic function, Eq. (12.5), by way of

$$\langle \lambda \rangle = \frac{1}{i} \frac{\partial}{\partial \theta} M(\theta) \big|_{\theta=0}. \qquad (12.28)$$

Using Eq. (12.7) we have,

$$\langle \lambda \rangle = \frac{1}{i} \frac{\partial}{\partial \theta} M(\theta) \big|_{\theta=0} = \frac{1}{i} \frac{\partial}{\partial \theta} \int \varphi^*(x) e^{i\theta A} \varphi(x) \, dx \big|_{\theta=0} \qquad (12.29)$$

or

$$\langle \lambda \rangle = \int \varphi^*(x) A \varphi(x) \, dx. \qquad (12.30)$$

Thus we have two dramatically different expressions for the expectation value, Eq. (12.27) and Eq. (12.30). Similarly for any function $f(\lambda)$,

$$\langle f(\lambda) \rangle = \int f(\lambda) |F(\lambda)|^2 d\lambda = \int \varphi^*(x) f(A) \varphi(x) \, dx. \qquad (12.31)$$

Quantum mechanics and the Born interpretation. In 1926, Born gave what is now accepted as the proper interpretation of the formalism of quantum mechanics, namely that it is probabilistic theory. The above method of calculating probabilities is what is called the Born interpretation and is one of the fundamental ideas of quantum mechanics. In quantum mechanics, observables are represented by operators and the numerical values that can be measured are the eigenvalues of the operators and the corresponding probability is given by Eq. (12.18) or Eq. (12.26).

12.3 Examples

Linear Combination of X and D. Consider the operator made up of a linear combination of X and D,

$$A = \alpha X + \beta D. \qquad (12.32)$$

The operator is Hermitian for real α and β. Solving the eigenvalue problem

$$\left(\alpha x - i\beta \frac{d}{dx} \right) u(\lambda, x) = \lambda u(\lambda, x) \qquad (12.33)$$

12.3. Examples

gives
$$u(\lambda, x) = \frac{1}{\sqrt{2\pi\beta}} e^{i(\lambda x - \alpha x^2/2)/\beta} \tag{12.34}$$

where we have normalized to a delta function. Hence, we have the following transform pairs:
$$F(\lambda) = \frac{1}{\sqrt{2\pi\beta}} \int \varphi(x) e^{-i(\lambda x - \alpha x^2/2)/\beta} dx, \tag{12.35}$$

$$\varphi(x) = \frac{1}{\sqrt{2\pi\beta}} \int F(\lambda) e^{i(\lambda x - \alpha x^2/2)/\beta} d\lambda. \tag{12.36}$$

For the characteristic function, using Eq. (12.7), we have
$$M(\theta) = \langle e^{i\theta A} \rangle \tag{12.37}$$

$$= \int \varphi^*(x) \, e^{i\theta(\alpha X + \beta D)} \varphi(x) \, dx \tag{12.38}$$

$$= \int \varphi^*(x) \, e^{i\theta^2 \alpha\beta/2} \, e^{i\alpha\theta X} \, e^{i\theta\beta D} \varphi(x) \, dx \tag{12.39}$$

$$= \int \varphi^*(x) \, e^{i\theta^2 \alpha\beta/2} \, e^{i\alpha\theta x} \varphi(x + \theta\beta) \, dx \tag{12.40}$$

giving
$$M(\theta) = \int \varphi^*(x - \tfrac{1}{2}\theta\beta) \, e^{i\theta\alpha x} \varphi(x + \tfrac{1}{2}\theta\beta) \, dx. \tag{12.41}$$

The distribution is then
$$P(\lambda) = \frac{1}{2\pi} \int M(\theta) \, e^{-i\theta\lambda} d\theta \tag{12.42}$$

$$= \frac{1}{2\pi} \iint \varphi^*(x - \tfrac{1}{2}\theta\beta) \, e^{-i\theta(\lambda - \alpha x)} \varphi(x + \tfrac{1}{2}\theta\beta) d\theta dx \tag{12.43}$$

which simplifies to
$$P(\lambda) = \left| \frac{1}{\sqrt{2\pi\beta}} \int \varphi(x) e^{-i(\lambda x - \alpha x^2/2)/\beta} dx \right|^2 = |F(\lambda)|^2 \tag{12.44}$$

and is consistent with the general result, Eq. (12.18).

Spatial Frequency Operator. If in the above we take $\alpha = 0$ and $\beta = 1$, then A is just the spatial frequency operator,
$$A = D. \tag{12.45}$$

From Eq. (12.34) we have that the eigenfunctions are
$$u(\lambda, x) = \frac{1}{\sqrt{2\pi}} e^{i\lambda x} \tag{12.46}$$

which of course leads to standard Fourier analysis. For the characteristic function we have

$$M(\theta) = \langle e^{i\theta D} \rangle = \int \varphi^*(x) e^{i\theta D} \varphi(x) \, dx \qquad (12.47)$$

$$= \int \varphi^*(x) \varphi(x + \theta) \, dx. \qquad (12.48)$$

The distribution is then

$$P(\lambda) = \frac{1}{2\pi} \int M(\theta) e^{-i\theta\lambda} d\theta = \frac{1}{2\pi} \iint \varphi^*(x) \varphi(x+\theta) e^{-i\theta\lambda} dx \, d\theta \qquad (12.49)$$

which simplifies to

$$P(\lambda) = \left| \frac{1}{\sqrt{2\pi}} \int \varphi(x) e^{-i\lambda x} dx \right|^2. \qquad (12.50)$$

In spectral analysis, this is called the energy density spectrum.

Example. $\alpha X + \beta D^2$. We solve the eigenvalue problem in the Fourier domain. We write

$$\{i\alpha \frac{d}{dp} + \beta p^2\} \, \widehat{u}(\lambda, p) = \lambda \, \widehat{u}(\lambda, p). \qquad (12.51)$$

The solutions, normalized to a delta function, are

$$\widehat{u}(\lambda, p) = \frac{1}{\sqrt{2\pi\alpha}} e^{-i(\lambda p - \beta p^3/3)/\alpha} \qquad (12.52)$$

and therefore for any state function in the Fourier representation $\widehat{\varphi}(p)$,

$$\widehat{\varphi}(p) = \frac{1}{\sqrt{2\pi\alpha}} \int \widehat{F}(\lambda) e^{-i(\lambda p - \beta p^3/3)/\alpha} d\lambda, \qquad (12.53)$$

$$\widehat{F}(\lambda) = \frac{1}{\sqrt{2\pi\alpha}} \int \widehat{\varphi}(p) e^{i(\lambda p - \beta p^3/3)/\alpha} dp. \qquad (12.54)$$

The eigenfunctions $\widehat{u}(\lambda, p)$ can be written in the position representation as

$$u(\lambda, x) = \frac{1}{\sqrt{2\pi}} \int \widehat{u}(\lambda, p) e^{ipx} dp \qquad (12.55)$$

$$= \frac{1}{2\pi} \frac{1}{\sqrt{\alpha}} \int e^{-i(\lambda p - \beta p^3/3)/\alpha} e^{ipx} dp \qquad (12.56)$$

which after some straightforward manipulations leads to

$$u(\lambda, x) = \frac{1}{\pi \alpha^{1/6} \beta^{1/3}} \int_0^\infty \cos\left[\frac{1}{3}p^3 + \frac{\alpha^{1/3}}{\beta^{1/3}}(x - \lambda/\alpha)p\right] dp. \qquad (12.57)$$

12.3. Examples

This can be written in terms of an Airy function which is defined by

$$\text{Ai}(x) = \frac{1}{\pi} \int_0^\infty \cos\left(\tfrac{1}{3}t^3 + xt\right) dt$$

and satisfies the differential equation

$$\frac{d^2}{dx^2}\text{Ai}(x) - x\text{Ai}(x) = 0. \tag{12.58}$$

Therefore we have that

$$u(\lambda, x) = \frac{1}{\alpha^{1/6}\beta^{1/3}} \text{Ai}\left(\frac{\alpha^{1/3}}{\beta^{1/3}}[x - \lambda/\alpha]\right). \tag{12.59}$$

Example. Scale Operator. Consider the operator

$$C = \tfrac{1}{2}(XD + DX) = XD - \tfrac{1}{2}i = DX + \tfrac{1}{2}i \tag{12.60}$$

which is sometimes called the dilation operator. The eigenvalue problem is

$$Cu(c, x) = c\,u(c, x) \tag{12.61}$$

or

$$\frac{1}{2i}\left(\frac{d}{dx}x + x\frac{d}{dx}\right) u(c, x) = c\,u(c, x). \tag{12.62}$$

The solutions normalized to a delta function are

$$u(c, x) = \frac{1}{\sqrt{2\pi}} \frac{e^{ic\ln x}}{\sqrt{x}}, \qquad x \geq 0 \tag{12.63}$$

and satisfy

$$\int_0^\infty u^*(c', x)u(c, x)\,dx = \delta(c - c'), \tag{12.64}$$

$$\int u^*(c, x')u(c, x)\,dc = \delta(x - x'), \qquad x, x' \geq 0. \tag{12.65}$$

Hence, the transformation equations between the x and c are

$$\varphi(x) = \frac{1}{\sqrt{2\pi}} \int F(c) \frac{e^{ic\ln x}}{\sqrt{x}} dc; \qquad x \geq 0, \tag{12.66}$$

$$F(c) = \frac{1}{\sqrt{2\pi}} \int_0^\infty \varphi(x) \frac{e^{-ic\ln x}}{\sqrt{x}} dx. \tag{12.67}$$

The operation of $e^{i\theta C}$ on an arbitrary function, $\varphi(x)$, is given by [19]

$$e^{i\theta C}\varphi(x) = e^{\theta/2}\varphi(e^\theta x) \tag{12.68}$$

and therefore the characteristic function is

$$M(\theta) = \int_0^\infty \varphi^*(x)\, e^{i\theta C}\, \varphi(x)\, dx = \int_0^\infty \varphi^*(x)\, e^{\theta/2}\, \varphi(e^\theta x)\, dx \qquad (12.69)$$

which simplifies to

$$M(\theta) = \int_0^\infty \varphi^*(e^{-\theta/2}x)\varphi(e^{\theta/2}x)\, dx. \qquad (12.70)$$

Hence

$$P(c) = \frac{1}{2\pi} \int \int_0^\infty \varphi^*(e^{-\theta/2}x)\varphi(e^{\theta/2}x)e^{-i\theta c}\, dx\, d\theta \qquad (12.71)$$

which reduces to

$$P(c) = \left| \frac{1}{\sqrt{2\pi}} \int_0^\infty \varphi(x) \frac{e^{-ic\ln x}}{\sqrt{x}} dx \right|^2 = |F(c)|^2. \qquad (12.72)$$

Example. Inverse Frequency. Consider the operator R defined by

$$R = \frac{p_0}{D} \qquad (12.73)$$

where p_0 is constant. We call R the inverse frequency operator and use r to signify the inverse frequency values. The eigenvalue problem in the Fourier representation is

$$\frac{p_0}{p} \widehat{u}(r, p) = r\, \widehat{u}(r, p). \qquad (12.74)$$

The solution, normalized to a delta function, is

$$\widehat{u}(r, p) = \frac{\sqrt{p_0}}{r} \delta(p - p_0/r) = \frac{\sqrt{p_0}}{p} \delta(r - p_0/p) \qquad (12.75)$$

and one can expand $\widehat{\varphi}(p)$ as

$$\widehat{\varphi}(p) = \int \widehat{F}(r)\widehat{u}(r, p)\, dr \qquad (12.76)$$

with

$$\widehat{F}(r) = \int \widehat{\varphi}(p)\widehat{u}^*(r, p)\, dp. \qquad (12.77)$$

Substituting Eq. (12.75) into Eq. (12.77) we obtain that

$$\widehat{F}(r) = \int \widehat{\varphi}(p) \frac{\sqrt{p_0}}{r} \delta(p - p_0/r)\, dp = \frac{\sqrt{p_0}}{r} \widehat{\varphi}(p_0/r). \qquad (12.78)$$

The distribution, according to Eq. (12.18), is then

$$P(r) = |F(r)|^2 = \frac{p_0}{r^2} |\widehat{\varphi}(p_0/r)|^2. \qquad (12.79)$$

12.3. Examples

It is of interest to calculate the characteristic function

$$M(\theta) = \int \widehat{\varphi}^*(p) e^{i\theta p_0/p} \widehat{\varphi}(p)\, dp = \int e^{i\theta p_0/p} |\widehat{\varphi}(p)|^2\, dp \qquad (12.80)$$

giving

$$P(r) = \frac{1}{2\pi} \iint e^{i\theta p_0/p} |\widehat{\varphi}(p)|^2 e^{-i\theta r}\, dp\, d\theta \qquad (12.81)$$

which evaluates to the $P(r)$ given by Eq. (12.79).

Chapter 13

Uncertainty Principle for Arbitrary Operators

The uncertainty principle was first given by Heisenberg and justified in a heuristic manner. The formulation in terms of standard deviations was derived by Weyl for the X, D case and by Robertson for the case of two arbitrary operators. The uncertainty principle involves the relationship of the standard deviation of two operators when calculated with the *same* state function. Since we are calculating the standard deviations with the same state function it is not surprising that there should be a relationship between the two standard deviations. But what is surprising is the form that the relationship takes and the fact that the commutator of the two operators appears in a fundamental way.

13.1 The Standard Deviation of an Operator

The average and average square of a Hermitian operator are, respectively, given by

$$\langle A \rangle = \int \varphi^*(x)\, A\, \varphi(x)\, dx, \tag{13.1}$$

$$\langle A^2 \rangle = \int \varphi^*(x)\, A^2\, \varphi(x)\, dx = \int |A\, \varphi(x)|^2\, dx, \tag{13.2}$$

and the variance is defined by

$$\sigma_A^2 = \langle A^2 \rangle - \langle A \rangle^2 \tag{13.3}$$

which can be written as

$$\sigma_A^2 = \int \varphi^*(x)\, (A - \langle A \rangle)^2 \varphi(x)\, dx = \int |(A - \langle A \rangle)\varphi(x)|^2\, dx. \tag{13.4}$$

Alternate expressions. Suppose we break up the quantity $A\varphi/\varphi$ into its real and imaginary parts,

$$\frac{A\varphi}{\varphi} = \left(\frac{A\varphi}{\varphi}\right)_R + i\left(\frac{A\varphi}{\varphi}\right)_I \tag{13.5}$$

and consider the expected value

$$\langle A \rangle = \int \varphi^*(x) A\varphi(x)\, dx \tag{13.6}$$

$$= \int \frac{A\varphi(x)}{\varphi(x)} |\varphi(x)|^2\, dx \tag{13.7}$$

$$= \int \left[\left(\frac{A\varphi}{\varphi}\right)_R + i\left(\frac{A\varphi}{\varphi}\right)_I\right] |\varphi(x)|^2\, dx. \tag{13.8}$$

But for Hermitian operators $\langle A \rangle$ is real and we therefore must have

$$\langle A \rangle = \int \left(\frac{A\varphi}{\varphi}\right)_R |\varphi(x)|^2\, dx. \tag{13.9}$$

Similarly, using Eq. (13.4) we have

$$\sigma_A^2 = \int \varphi^*(x)\,(A - \langle A \rangle)^2\, \varphi(x)\, dx \tag{13.10}$$

$$= \int |\,\{A - \langle A \rangle\}\varphi(x)\,|^2\, dx \tag{13.11}$$

$$= \int \left|\left(\frac{A\varphi}{\varphi} - \langle A \rangle\right)\varphi(x)\right|^2 dx \tag{13.12}$$

$$= \int \left|\left[\left(\frac{A\varphi}{\varphi}\right)_R - \langle A \rangle + i\left(\frac{A\varphi}{\varphi}\right)_I\right]\varphi(x)\right|^2 dx \tag{13.13}$$

and therefore

$$\sigma_A^2 = \int \left(\frac{A\varphi}{\varphi}\right)_I^2 |\varphi(x)|^2\, dx + \int \left[\left(\frac{A\varphi}{\varphi}\right)_R - \langle A \rangle\right]^2 |\varphi(x)|^2\, dx. \tag{13.14}$$

Similarly,

$$\langle A^2 \rangle = \int \left\{\left(\frac{A\varphi}{\varphi}\right)_I^2 + \left(\frac{A\varphi}{\varphi}\right)_R^2\right\} |\varphi(x)|^2\, dx. \tag{13.15}$$

It is important to appreciate that the above results are based on taking the operators to be Hermitian.

Example. Consider the case where we take the state function to be

$$\varphi(x) = (\alpha/\pi)^{1/4}\, e^{-\alpha x^2/2 + i\beta x^2/2 + i\gamma x} \tag{13.16}$$

and we consider $\langle D \rangle$ and σ_D^2. We have

$$\frac{D\varphi(x)}{\varphi(x)} = -i(-\alpha x + i\beta x + i\gamma) \tag{13.17}$$

and therefore

$$\left(\frac{D\varphi}{\varphi}\right)_R = \beta x + \gamma; \qquad \left(\frac{D\varphi}{\varphi}\right)_I = \alpha x. \tag{13.18}$$

Using Eq. (13.9) we have

$$\langle D \rangle = \int (\beta x + \gamma) \, |\varphi(x)|^2 \, dx = (\alpha/\pi)^{1/2} \int (\beta x + \gamma) e^{-\alpha x^2} = \gamma. \tag{13.19}$$

Now consider

$$\sigma_D^2 = \int \left(\frac{D\varphi}{\varphi}\right)_I^2 |\varphi(x)|^2 \, dx + \int \left[\left(\frac{D\varphi}{\varphi}\right)_R - \langle D \rangle\right]^2 |\varphi(x)|^2 \, dx \tag{13.20}$$

$$= \int (\alpha x)^2 \, |\varphi(x)|^2 \, dx + \int [\,\beta x + \gamma - \gamma\,]^2 |\varphi(x)|^2 \, dx \tag{13.21}$$

$$= (\alpha/\pi)^{1/2}(\alpha^2 + \beta^2) \int x^2 \, e^{-\alpha x^2} \, dx \tag{13.22}$$

giving

$$\sigma_D^2 = \frac{\alpha^2 + \beta^2}{2\alpha}. \tag{13.23}$$

13.2 Schwarz Inequality

The Schwarz inequality is that for any two complex functions $f(x)$ and $g(x)$

$$\int |f(x)|^2 \, dx \times \int |g(x)|^2 \, dx \geq \left|\int f^*(x) g(x) \, dx\right|^2. \tag{13.24}$$

A simple proof is to use the identity

$$\int |f(x)|^2 \, dx \int |g(x)|^2 \, dx - \left|\int f^*(x) g(x) \, dx\right|^2$$

$$= \tfrac{1}{2}\iint |f(x) g(y) - f(y) g(x)|^2 \, dx \, dy \tag{13.25}$$

which is readily verified by direct expansion. Since the right hand side is manifestly positive, Eq. (13.24) follows.

13.3 Uncertainty Principle

The uncertainty principle involves the standard deviation of two operators. Since the standard deviation does not depend on the average, there is no loss of generality if we consider operators whose means are zero. For two such operators, A and B, the standard deviations are

$$\sigma_A^2 = \langle A^2 \rangle = \int |A\varphi(x)|^2 \, dx, \tag{13.26}$$

$$\sigma_B^2 = \langle B^2 \rangle = \int |B\varphi(x)|^2 \, dx. \tag{13.27}$$

Multiply these two expressions

$$\sigma_A^2 \sigma_B^2 = \int |A\varphi(x)|^2 \, dx \times \int |B\varphi(x)|^2 \, dx. \tag{13.28}$$

Letting

$$f(x) = A\varphi(x), \tag{13.29}$$
$$g(x) = B\varphi(x), \tag{13.30}$$

and using the Schwarz inequality, Eq. (13.24), we have

$$\sigma_A^2 \sigma_B^2 = \int |A\varphi(x)|^2 \, dx \times \int |B\varphi(x)|^2 \, dx \geq \left| \int [A\varphi(x)]^* B\varphi(x) \, dx \right|^2 \tag{13.31}$$

$$= \left| \int \varphi^*(x) A B \varphi(x) \, dx \right|^2. \tag{13.32}$$

That is

$$\sigma_A^2 \sigma_B^2 \geq |\langle AB \rangle|^2. \tag{13.33}$$

Now, AB is not Hermitian and therefore the expected value is not real, but using Eq. (2.17) we can write

$$AB = \tfrac{1}{2}[A, B]_+ + \tfrac{i}{2}[A, B]/i \tag{13.34}$$

where $\tfrac{1}{2}[A, B]_+$ and $\tfrac{1}{2i}[A, B]$ are Hermitian. Therefore both $\langle \tfrac{1}{2}[A, B]_+ \rangle$ and $\langle \tfrac{1}{2i}[A, B] \rangle$ are real and hence,

$$|\langle AB \rangle|^2 = |\langle \tfrac{1}{2}[A, B]_+ + i\tfrac{1}{2i}[A, B] \rangle|^2 \tag{13.35}$$

$$= \tfrac{1}{4}|\langle [A, B]_+ \rangle|^2 + \tfrac{1}{4}|\langle [A, B] \rangle|^2. \tag{13.36}$$

Therefore

$$\sigma_A^2 \sigma_B^2 \geq \tfrac{1}{4}|\langle [A, B]_+ \rangle|^2 + \tfrac{1}{4}|\langle [A, B] \rangle|^2 \quad \text{(second UP)}. \tag{13.37}$$

If the first term is dropped we have

$$\sigma_A^2 \sigma_B^2 \geq \tfrac{1}{4}|\langle[A,B]\rangle|^2 \tag{13.38}$$

or

$$\sigma_A \sigma_B \geq \tfrac{1}{2}|\langle[A,B]\rangle| \quad \text{(first UP)}. \tag{13.39}$$

We have called Eq. (13.39) the first uncertainty principle because that is the most commonly used one. Eq. (13.37), the second uncertainty principle, is of course stronger.

The equation that minimizes the uncertainty product, $\sigma_A \sigma_B$, is given by

$$(B - \langle B \rangle)\,\varphi(x) = \lambda(A - \langle A \rangle)\,\varphi(x) \tag{13.40}$$

where

$$\lambda = \frac{\langle[A,B]\rangle}{2\sigma_A^2}. \tag{13.41}$$

Note that in general λ may be complex since $[A,B]$ is not Hermitian.

It is important to appreciate the following. The right hand side of the uncertainty principle in, $\tfrac{1}{2}|\langle[A,B]\rangle|$, is in general state dependent. The only case for which it is not is when $[A,B]$ is a constant. However, even in that case, if we use the uncertainty principle as given by Eq. (13.37) then it is state dependent. The example below will illustrate the issue.

The (X,D) Case. For the X and D case, $[X,D] = i$, and therefore

$$\sigma_X \sigma_D \geq \tfrac{1}{2} \quad \text{(first UP)}. \tag{13.42}$$

This is one of the most renowned results both in quantum mechanics and signal analysis.

13.4 Examples

Example 1. Consider the uncertainty principle for X and D where the state function is given by Eq. (13.16),

$$\varphi(x) = (\alpha/\pi)^{1/4} e^{-\alpha x^2/2 + i\beta x^2/2 + i\gamma x}. \tag{13.43}$$

We first give the exact value for $\sigma_X \sigma_D$. It is straightforward to verify that

$$\langle X \rangle =, \qquad \sigma_X^2 = \frac{1}{2\alpha}, \tag{13.44}$$

$$\langle D \rangle = \gamma, \qquad \langle D^2 \rangle = \frac{\alpha^2 + \beta^2}{2\alpha} + \gamma^2, \tag{13.45}$$

$$\sigma_D^2 = \frac{\alpha^2 + \beta^2}{2\alpha}. \tag{13.46}$$

Therefore
$$\sigma_X \sigma_D = \tfrac{1}{2}\sqrt{1 + \tfrac{\beta^2}{\alpha^2}} \qquad \text{(exact)} \qquad (13.47)$$
and indeed $\sigma_X \sigma_D \geq \tfrac{1}{2}$.

The spectrum of $\varphi(x)$ is calculated to be
$$\widehat{\varphi}(p) = \sqrt{\frac{\sqrt{\alpha}}{\sqrt{\pi}(\alpha - i\beta)}}\, e^{-(p-\gamma)^2/2(\alpha - i\beta)}. \qquad (13.48)$$

Now consider $|\widehat{\varphi}(p)|^2$ and $|\varphi(x)|^2$,
$$|\varphi(x)|^2 = (\alpha/\pi)^{1/2} e^{-\alpha x^2}, \qquad (13.49)$$
$$|\widehat{\varphi}(p)|^2 = \sqrt{\frac{\alpha}{\pi(\alpha^2 + \beta^2)}}\, e^{-\alpha(p-\gamma)^2/(\alpha^2 + \beta^2)}, \qquad (13.50)$$

and we see that no matter how we adjust α and β, $|\widehat{\varphi}(p)|^2$ and $|\varphi(x)|^2$ can not both be made arbitrarily narrow. The uncertainty principle is a reflection of this fact.

Now consider the uncertainty principle as given by Eq. (13.37). We have to calculate $\langle [X, D]_+ \rangle$. Using the fact that
$$XD + DX = XD + XD - i = 2XD - i \qquad (13.51)$$
we have
$$\langle [X, D]_+ \rangle = \int \varphi^*(x) \left(2x \frac{d}{i\,dx} - i \right) \varphi(x) \qquad (13.52)$$
$$= -i (\alpha/\pi)^{1/2} \int [2x(-\alpha x + i\beta x + i\gamma) + 1] e^{-\alpha x^2} dx \qquad (13.53)$$

which evaluates to
$$\langle [X, D]_+ \rangle = \frac{\beta}{\alpha}. \qquad (13.54)$$
Therefore if we use the uncertainty principle given by Eq. (13.37) we have
$$\sigma_X^2 \sigma_D^2 \geq \tfrac{1}{4} |\langle [X, D]_+ \rangle|^2 + \tfrac{1}{4} |\langle [X, D] \rangle|^2 \qquad (13.55)$$
$$= \frac{1}{4}\left(1 + \frac{\beta^2}{\alpha^2}\right) \qquad \text{(second UP)} \qquad (13.56)$$
which is the exact answer.

Example. X and D^2. We now consider the uncertainty between X and D^2 for the state function given by Eq. (13.16). The commutator is
$$[X, D^2] = 2Di \qquad (13.57)$$

13.4. Examples

and therefore
$$\sigma_X^2 \sigma_{D^2}^2 \geq |\langle D \rangle|^2 = \gamma^2 \qquad \text{(first UP)}. \tag{13.58}$$

Notice that if we take $\gamma = 0$ then $\sigma_X^2 \sigma_{D^2}^2 \geq 0$ which is vacuous. We now calculate the exact $\sigma_X^2 \sigma_{D^2}^2$. Straightforward calculation yields

$$\langle D^2 \rangle = \sigma_D^2 + \gamma^2, \tag{13.59}$$
$$\langle D^4 \rangle = \gamma^4 + 6\gamma^2 \sigma_D^2 + 3\sigma_D^4, \tag{13.60}$$

where σ_D is given by Eq. (13.46). Therefore

$$\sigma_{D^2}^2 = \langle D^4 \rangle - \langle D^2 \rangle^2 = 4\gamma^2 \sigma_D^2 + 2\sigma_D^4 = 2 \left(\frac{\alpha^2 + \beta^2}{\alpha} \right) \left(\frac{\alpha^2 + \beta^2}{4\alpha} + \gamma^2 \right) \quad \text{(exact)} \tag{13.61}$$

giving
$$\sigma_X^2 \sigma_{D^2}^2 = \left(1 + \frac{\beta^2}{\alpha^2} \right) \left(\frac{\alpha^2 + \beta^2}{4\alpha} + \gamma^2 \right) \qquad \text{exact} \tag{13.62}$$

for the exact uncertainty product. To reconcile with Eq. (13.58), the first term can be dropped since it is manifestly positive

$$\sigma_X^2 \sigma_{D^2}^2 \geq \left(1 + \frac{\beta^2}{\alpha^2} \right) \gamma^2 \geq \gamma^2. \tag{13.63}$$

Now consider the second uncertainty principle. We have that

$$[X, D^2]_+ = 2XD^2 - 2Di = 2D^2 X + 2Di \tag{13.64}$$

and direct calculation gives

$$\langle [X, D^2]_+ \rangle = \frac{2\beta}{\alpha} \gamma. \tag{13.65}$$

Hence for the second uncertainty principle Eq. (13.37), we obtain

$$\sigma_X^2 \sigma_{D^2}^2 \geq \tfrac{1}{4} |\langle [A, B]_+ \rangle|^2 + \tfrac{1}{4} |\langle [A, B] \rangle|^2 = \left(\frac{\beta}{\alpha} \gamma \right)^2 + \gamma^2 \tag{13.66}$$

or
$$\sigma_X^2 \sigma_{D^2}^2 \geq \left(1 + \frac{\beta^2}{\alpha^2} \right) \gamma^2 \qquad \text{(second UP)}. \tag{13.67}$$

In comparing with the exact answer, Eq. (13.62), we see that the second uncertainty principle is considerably stronger than the first.

Example. Consider now the special case where $\beta = \gamma = 0$. From Eq. (13.62) we have

$$\sigma_X^2 = \frac{1}{2\alpha}, \qquad \sigma_{D^2}^2 = \frac{\alpha^2}{2} \tag{13.68}$$

and hence
$$\sigma_X^2 \sigma_{D^2}^2 = \frac{\alpha}{4} \quad \text{(exact)}. \tag{13.69}$$

Note that now the uncertainty product can be made as small as one desires by controlling α. What is happening is that for $\alpha \to 0$ the standard deviation of D^2 is going to zero faster than the standard deviation of x is going to infinity.

It is of interest to calculate the intensity functions. For x it is given by Eq. (13.49)
$$I(x) = |\varphi(x)|^2 = (\alpha/\pi)^{1/2} e^{-\alpha x^2} \tag{13.70}$$

where we have used $I(x)$ for the intensity to differentiate the examples. Now for D^2 we use ν for the new variable corresponding to D^2,
$$\nu = p^2 \tag{13.71}$$

and the intensity, $I(\nu)$ is obtained by way of Eq. (13.49)
$$\nu = p^2, \tag{13.72}$$
$$I(\nu) = \frac{1}{\sqrt{\nu}} |\widehat{\varphi}(\sqrt{\nu})|^2 = \sqrt{\frac{1}{\pi \alpha \nu}} e^{-\nu/\alpha}, \tag{13.73}$$

where $|\widehat{\varphi}(p)|^2$ is given by Eq. (13.50). The intensity, $I(\nu)$, is narrower as compared to $I(x)$. One may calculate the standard deviation of $I(\nu)$ directly:
$$\sigma_\nu^2 = \int_0^\infty (\nu - \langle \nu \rangle)^2 I(\nu) d\nu = \sigma_{D^2}^2 \tag{13.74}$$

and obtain $\sigma_{D^2}^2$ as given by Eq. (13.68).

Chapter 14

The Khintchine Theorem and Characteristic Function Representability

In the standard consideration of the characteristic function, defined by the Fourier transform of the probability density, there arises the issue that not every complex function is a characteristic function since it must be derivable from a proper probability distribution. That is, the characteristic function is the Fourier transform of a function that is manifestly positive. Khintchine derived necessary and sufficient conditions for a complex function to be a characteristic function: $M(\theta)$ is a characteristic function if and only if it is expressible as

$$M(\theta) = \int \varphi^*(x)\varphi(x+\theta)\,dx \qquad (14.1)$$

for some function $\varphi(x)$, which must be normalized to 1,

$$\int |\varphi(x)|^2\,dx = 1. \qquad (14.2)$$

One then says that $M(\theta)$ is representable or realizable.

Now, if we write Eq. (14.1) as

$$M(\theta) = \int \varphi^*(x)e^{i\theta D}\varphi(x)\,dx \qquad (14.3)$$

we see that it is a special case of the result that we obtained in Chap. 12, Eq. (12.7), where we showed that

$$M(\theta) = \int \varphi^*(x)e^{i\theta A}\varphi(x)\,dx \qquad (14.4)$$

is a proper characteristic function if the operator A is Hermitian. The associated probability distribution is, for the continuous case, given by

$$P(\lambda) = |F(\lambda)|^2 = \left| \int \varphi(x) u^*(\lambda, x) \, dx \right|^2 \tag{14.5}$$

where $u(\lambda, x)$ are the eigenfunctions of the operator A. Thus, the Khintchine theorem is a special case where the operator is D. One is led to generalizing the Khintchine theorem: $M(\theta)$ is a characteristic function if and only if there exists the representation

$$M(\theta) = \int \varphi^*(x) e^{i\theta A} \varphi(x) \, dx \tag{14.6}$$

where A is a Hermitian operator.

Since Eq. (14.5) is manifestly positive, the sufficiency follows. Now suppose we have the probability distribution $P(\lambda)$, then by the definition of $M(\theta)$,

$$M(\theta) = \int e^{i\theta\lambda} P(\lambda) \, d\lambda = \int e^{i\theta\lambda} \sqrt{P(\lambda)} \sqrt{P(\lambda)} \, d\lambda. \tag{14.7}$$

Expand $\sqrt{P(\lambda)}$ as

$$\sqrt{P(\lambda)} = \int u^*(\lambda, x) \, \varphi(x) \, dx. \tag{14.8}$$

But since $\sqrt{P(\lambda)}$ is real we also have

$$\sqrt{P(\lambda)} = \int u(\lambda, x) \, \varphi^*(x) \, dx. \tag{14.9}$$

Substituting into Eq. (14.7), gives

$$M(\theta) = \iiint u^*(\lambda, x') \varphi(x') e^{i\theta\lambda} u(\lambda, x) \varphi^*(x) \, dx \, dx' \, d\lambda \tag{14.10}$$

$$= \iiint u^*(\lambda, x') \varphi(x') \left[e^{i\theta A} u(\lambda, x) \right] \varphi^*(x) \, dx \, dx' \, d\lambda \tag{14.11}$$

$$= \iint \varphi(x') \left[e^{i\theta A} \delta(x' - x) \right] \varphi^*(x) \, dx \, dx' \tag{14.12}$$

$$= \int \varphi^*(x) e^{i\theta A} \varphi(x) \, dx \tag{14.13}$$

which proves that Eq. (14.4) is a necessary condition. One can prove the analogous results for the discrete case.

Chapter 15

Arbitrary operators: Two Operators

In the previous chapters, we studied the association of the variables (x, p) with an operator where the individual correspondences of x and p are associated with the operators X and D. In this chapter, we consider the generalization to arbitrary operators [21, 22, 79]. A theory analogous to the (X, D) case has not been developed for arbitrary operators but special cases have been studied and we discuss some of these. To keep the distinction between the (X, D) and general case, we use $\sigma(\alpha, \beta)$ and $G(A,B)$ for the symbol and corresponding operator, respectively.

15.1 Operator Association

Suppose the ordinary variables α, β are associated with the operators A and B by

$$A \leftrightarrow \alpha, \qquad (15.1)$$
$$B \leftrightarrow \beta, \qquad (15.2)$$

then we want to associate the symbol $\sigma(\alpha, \beta)$ with an operator $G(A, B)$,

$$G(A,B) \leftrightarrow \sigma(\alpha, \beta). \qquad (15.3)$$

In the most general case, the operators A and B are arbitrary Hermitian operators with commutator, C,

$$[A,B] = C. \qquad (15.4)$$

As with the (x, p) case, we express the symbol in terms of its Fourier transform,

$$\widehat{\sigma}(\theta, \tau) = \iint \sigma(\alpha, \beta) e^{-i\theta\alpha - i\tau\beta} \, d\alpha \, d\beta, \qquad (15.5)$$

$$\sigma(\alpha,\beta) = \frac{1}{4\pi^2} \iint \hat{\sigma}(\theta,\tau) e^{i\theta\alpha+i\tau\beta} \, d\theta \, d\tau. \qquad (15.6)$$

One possibility for the operator $G(A,B)$ is to replace $e^{i\theta\alpha+i\tau\beta}$ by $e^{i\theta A+i\tau B}$ in Eq. (15.6)

$$G(A,B) = \iint \hat{\sigma}(\theta,\tau) e^{i\theta A+i\tau B} \, d\theta \, d\tau. \qquad (15.7)$$

However, as for the (X, D) case there are an infinite number of other possibilities and in this chapter we consider three choices,

$$e^{i\theta\alpha+i\tau\beta} \leftrightarrow \begin{cases} e^{i\theta A+i\tau B} &= \mathcal{M}_1(\theta,\tau), \\ e^{i\theta A} e^{i\tau B} &= \mathcal{M}_2(\theta,\tau), \\ e^{i\theta A/2} e^{i\tau B} e^{i\theta A/2} &= \mathcal{M}_3(\theta,\tau), \end{cases} \qquad (15.8)$$

and we write

$$G(A,B) = \iint \hat{\sigma}(\theta,\tau) \, \mathcal{M}(\theta,\tau) \, d\theta \, d\tau \qquad (15.9)$$

where $\mathcal{M}(\theta,\tau)$ is one of the possible orderings, examples being the three choices given by Eq. (15.8). For the operation on a function we write

$$G(A,B)\varphi(x) = \iint \hat{\sigma}(\theta,\tau) \, \mathcal{M}(\theta,\tau)\varphi(x) \, d\theta \, d\tau. \qquad (15.10)$$

The general case can be written as

$$G^{\Phi}(A,B) = \iint \Phi(\theta,\tau)\hat{\sigma}(\theta,\tau) e^{i\theta A+i\tau B} \, d\theta \, d\tau \qquad (15.11)$$

where $\Phi(\theta,\tau)$ is the kernel defined in Chap. 5. One can put constraints on the kernel, as was done for the (x,p) case. For example, if we want

$$f(A) \leftrightarrow f(\alpha), \qquad (15.12)$$
$$g(B) \leftrightarrow g(\beta), \qquad (15.13)$$

where f and g are one variable functions, the constraints on the kernel are the same as for the (X, D) case, namely $\Phi(\theta,0) = \Phi(0,\tau) = 1$. However, as mentioned, a complete theory for Eq. (15.11) has not been developed.

15.2 Joint Distribution

As with the standard case, one can develop the concept of joint distributions in the sense that we seek a $P(\alpha,\beta)$ so that for an arbitrary state, $\varphi(x)$, we have

$$\int \varphi^*(x) \, G(A,B) \, \varphi(x) \, dx = \iint \sigma(\alpha,\beta) \, P(\alpha,\beta) \, d\alpha \, d\beta. \qquad (15.14)$$

15.3. Commuting Operators

We recall that the characteristic function is defined by

$$M(\theta,\tau) = \langle e^{i\theta\alpha+i\tau\beta}\rangle = \iint e^{i\theta\alpha+i\tau\beta}\, P(\alpha,\beta)\, d\alpha\, d\beta \tag{15.15}$$

and that the distribution is given by

$$P(\alpha,\beta) = \frac{1}{4\pi^2}\iint M(\theta,\tau)e^{-i\theta\alpha-i\tau\beta}\, d\theta\, d\tau. \tag{15.16}$$

As in the (X,D) case, the characteristic function is an average, the average of $\mathcal{M}(\theta,\tau)$, and we calculate it by way of

$$M(\theta,\tau) = \langle \mathcal{M}(\theta,\tau)\rangle = \int \varphi^*(x)\,\mathcal{M}(\theta,\tau)\,\varphi(x)\, dx. \tag{15.17}$$

The distribution is therefore

$$P(\alpha,\beta) = \frac{1}{4\pi^2}\iint \langle\mathcal{M}(\theta,\tau)\rangle\, e^{-i\theta\alpha-i\tau\beta}\, d\theta\, d\tau. \tag{15.18}$$

Once a particular choice is explicitly obtained, an infinite number of others can be obtained from

$$P_\Phi(\alpha,\beta) = \frac{1}{4\pi^2}\iint \Phi(\theta,\tau)\mathcal{M}(\theta,\tau)\, e^{-i\theta\alpha-i\tau\beta}\, d\theta\, d\tau. \tag{15.19}$$

Eq. (15.19) is the generalization of Eq. (5.2) for arbitrary operators.

15.3 Commuting Operators

If the operators commute,

$$[A,B] = 0 \tag{15.20}$$

then the association is unambiguously given by

$$e^{i\theta\alpha+i\tau\beta} \leftrightarrow e^{i\theta A}\, e^{i\tau B} = e^{i\tau B}e^{i\theta A} \tag{15.21}$$

and the corresponding operator for a symbol is given by

$$G(A,B) = \iint \hat{\sigma}(\theta,\tau)\, e^{i\theta A}\, e^{i\tau B}\, d\theta\, d\tau = \iint \hat{\sigma}(\theta,\tau)\, e^{i\tau B}\, e^{i\theta A} d\theta\, d\tau. \tag{15.22}$$

Commuting operators have common eigenfunctions and we may write

$$Au(\lambda,x) = \lambda\, u(\lambda,x) \tag{15.23}$$
$$Bu(\lambda,x) = b(\lambda)\, u(\lambda,x) \tag{15.24}$$

where λ and $b(\lambda)$ are the eigenvalues of A and B respectively and $u(\lambda, x)$ are the common eigenfunctions. We expand the state function as

$$\varphi(x) = \int u(\lambda, x) F(\lambda) \, d\lambda \qquad (15.25)$$

with

$$F(\lambda) = \int u^*(\lambda, x) \, \varphi(x) \, dx \qquad (15.26)$$

in which case we have

$$G(A,B)\varphi(x) = \iint \widehat{\sigma}(\theta, \tau) \, e^{i\theta A} \, e^{i\tau B} \, \varphi(x) \, d\theta \, d\tau \qquad (15.27)$$

$$= \iiint \widehat{\sigma}(\theta, \tau) \, e^{i\theta \lambda} \, e^{i\tau b(\lambda)} \, u(\lambda, x) F(\lambda) \, d\theta \, d\tau \, d\lambda \qquad (15.28)$$

$$= \frac{1}{4\pi^2} \iiint \sigma(\alpha, \beta) e^{-i\theta\alpha - i\tau\beta} e^{i\theta\lambda} \, e^{i\tau b(\lambda)} \, u(\lambda, x) F(\lambda) \, d\theta \, d\tau \, d\lambda \qquad (15.29)$$

giving

$$G(A,B)\varphi(x) = \int \sigma(\lambda, b(\lambda)) u(\lambda, x) F(\lambda) \, d\lambda. \qquad (15.30)$$

The expectation value of $G(A,B)$ is

$$\langle G(A,B) \rangle = \int \varphi^*(x) \, G(A,B) \, \varphi(x) \, dx \qquad (15.31)$$

$$= \iint u^*(\lambda', x) F^*(\lambda') \sigma(\lambda, b(\lambda)) u(\lambda, x) F(\lambda) \, d\lambda \, d\lambda' \, dx \qquad (15.32)$$

which simplifies to

$$\langle G(A,B) \rangle = \int \sigma(\lambda, b(\lambda)) \, |F(\lambda)|^2 \, d\lambda. \qquad (15.33)$$

To satisfy Eq. (15.14), we must find $P(\alpha, \beta)$ so that

$$\int \sigma(\lambda, b(\lambda)) \, |F(\lambda)|^2 \, d\lambda = \iint \sigma(\alpha, \beta) \, P(\alpha, \beta) \, d\alpha \, d\beta. \qquad (15.34)$$

Clearly, we must take

$$P(\alpha, \beta) = \delta(\beta - b(\alpha)) \, |F(\alpha)|^2. \qquad (15.35)$$

This distribution is manifestly positive and shows that for commuting operators, the procedure gives a proper distribution. Moreover, it brings in the important physical quantities, namely the eigenvalues of the operators.

15.4. The $e^{i\theta A+i\tau B}$ Correspondence

Characteristic function approach. We now obtain the same result from the point of view of the characteristic function. Using Eq. (15.21), we have

$$M(\theta, \tau) = \int \varphi^*(x)\, e^{i\theta A}\, e^{i\tau B}\, \varphi(x)\, dx \tag{15.36}$$

$$= \int u^*(\lambda', x) F^*(\lambda') e^{i\theta \lambda}\, e^{i\tau b(\lambda)}\, u(\lambda, x) F(\lambda)\, d\lambda\, d\lambda'\, dx \tag{15.37}$$

which evaluates to

$$M(\theta, \tau) = \int e^{i\theta \lambda}\, e^{i\tau b(\lambda)}\, |F(\lambda)|^2\, d\lambda. \tag{15.38}$$

The distribution is therefore

$$P(\alpha, \beta) = \frac{1}{4\pi^2} \iint \langle M(A, B) \rangle\, e^{-i\theta\alpha - i\tau\beta}\, d\theta\, d\tau \tag{15.39}$$

$$= \frac{1}{4\pi^2} \iint e^{i\theta\lambda}\, e^{i\tau b(\lambda)}\, |F(\lambda)|^2\, e^{-i\theta\alpha - i\tau\beta}\, d\lambda\, d\theta\, d\tau \tag{15.40}$$

$$= \int \delta(\lambda - \alpha)\delta(b(\lambda) - \beta)\, |F(\lambda)|^2\, d\lambda \tag{15.41}$$

$$= \delta(b(\alpha) - \beta)\, |F(\alpha)|^2 \tag{15.42}$$

which is the same as Eq. (15.35).

15.4 The $e^{i\theta A+i\tau B}$ Correspondence

In this section we study the correspondence $e^{i\theta A+i\tau B}$, where we take

$$G(A, B) = \iint \hat{\sigma}(\theta, \tau) e^{i\theta A + i\tau B}\, d\theta\, d\tau \tag{15.43}$$

and for the operation on a function we write

$$G(A, B)\varphi(x) = \iint \hat{\sigma}(\theta, \tau) e^{i\theta A + i\tau B} \varphi(x)\, d\theta\, d\tau. \tag{15.44}$$

For a state $\varphi(x)$, the characteristic function is

$$M(\theta, \tau) = \left\langle e^{i\theta A + i\tau B} \right\rangle = \int \varphi^*(x)\, e^{i\theta A + i\tau B}\, \varphi(x)\, dx \tag{15.45}$$

and the distribution is therefore

$$P(\alpha, \beta) = \frac{1}{4\pi^2} \iiint e^{-i\theta\alpha - i\tau\beta} \varphi^*(x) e^{i\theta A + i\tau B}\, \varphi(x)\, d\theta\, d\tau\, dx. \tag{15.46}$$

This ensures that

$$\int \varphi^*(x) G(A, B) \varphi(x)\, dx = \iint \sigma(\alpha, \beta) P(\alpha, \beta)\, d\alpha\, d\beta. \tag{15.47}$$

15.4.1 Disentanglement of $e^{i\theta A + i\tau B}$

Carrying out the operation $e^{i\theta A+i\tau B}\varphi(x)$ is generally very difficult. Equivalently, the problem of the disentanglement of $e^{i\theta A+i\tau B}$ is also difficult. Many general expansions have been given, among them the most important are the Zassenhaus formula

$$e^{A+B} = e^A e^B e^{-\frac{1}{2}[A,B]} e^{\frac{1}{3}[B,[A,B]] + \frac{1}{6}[A,[A,B]]} \tag{15.48}$$

$$\times e^{-\frac{1}{24}([[[A,B],A],A] + 3[[[A,B],A],B] + 3[[[A,B],B],B])} \cdots \tag{15.49}$$

and Baker-Cambell-Hausdorf formula

$$e^A e^B = e^{A+B+\frac{1}{2}[A,B] + \frac{1}{12}\{[A,[A,B]] - [B,[A,B]]\} - \frac{1}{24}[B,[A,[A,B]]] \cdots}. \tag{15.50}$$

These formulas are sometimes helpful when the series terminates. We also mention the Lie-Trotter product formula

$$e^{A+B} = \lim_{N \to \infty} (e^{A/N} e^{B/N})^N. \tag{15.51}$$

Special cases. There are cases where the disentanglement can be done explicitly. We list two important ones. The first is when

$$[A,[A,B]] = [B,[A,B]] = 0 \tag{15.52}$$

in which case

$$e^{i\theta A + i\tau B} = e^{i\theta A} e^{i\tau B} e^{\frac{1}{2}\theta\tau[A,B]} = e^{i\tau B} e^{i\theta A} e^{-\frac{1}{2}\theta\tau[A,B]}. \tag{15.53}$$

Hence

$$G(A,B) = \iint \hat{\sigma}(\theta,\tau) e^{-\frac{1}{2}\theta\tau[A,B]} e^{i\tau B} e^{i\theta A} \, d\theta \, d\tau \tag{15.54}$$

$$= \iint \hat{\sigma}(\theta,\tau) e^{\frac{1}{2}\theta\tau[A,B]} e^{i\theta A} e^{i\tau B} \, d\theta \, d\tau \tag{15.55}$$

and the operation on a function is then

$$G(A,B)\varphi(x) = \iint \hat{\sigma}(\theta,\tau) e^{\frac{1}{2}\theta\tau[A,B]} e^{i\tau A} e^{i\theta B} \varphi(x) \, d\theta \, d\tau. \tag{15.56}$$

The second is

$$e^{i\theta A + i\tau B} = e^{i\mu\theta\gamma/\xi} e^{i\theta\mu A} e^{i\tau B} e^{i\theta A} \quad \text{if } [A,B] = \gamma + \xi A \tag{15.57}$$

where

$$\mu = \frac{1}{i\tau\xi}\left[1 - (1 + i\tau\xi)e^{-i\tau\xi}\right]. \tag{15.58}$$

The corresponding operator association is therefore

$$G(A,B) = \iint \hat{\sigma}(\theta,\tau) e^{i\mu\theta\gamma/\xi} e^{i\theta\mu A} e^{i\tau B} e^{i\theta A} \, d\theta \, d\tau. \tag{15.59}$$

15.4.2 General procedure for evaluating $e^{i\theta A+i\tau B}\varphi(x)$

We now give a procedure for the evaluation of $e^{i\theta A+i\tau B}\varphi(x)$ that works in principle and often works in practice. It is similar to the procedure described in Sec. 2.10 for the operators X and D. Consider the operator $\theta A+\tau B$; since θ and τ are real the operator is Hermitian and hence the eigenfunctions are complete and orthogonal. For the eigenvalue problem we write,

$$\{\theta A + \tau B\}\, u(\lambda, x) = \lambda\, u(\lambda, x) \tag{15.60}$$

where λ and $u(\lambda, x)$ are the eigenvalues and eigenfunctions respectively. We address the continuous case first. Any function, $\varphi(x)$, can be expanded as

$$\varphi(x) = \int u(\lambda, x) F(\lambda)\, d\lambda \tag{15.61}$$

with

$$F(\lambda) = \int u^*(\lambda, x)\varphi(x)\, dx. \tag{15.62}$$

Now

$$e^{i\theta A+i\tau B}\varphi(x) = \int e^{i\lambda} u(\lambda, x) F(\lambda)\, d\lambda \tag{15.63}$$

and substituting for $F(\lambda)$ we have

$$e^{i\theta A+i\tau B}\varphi(x) = \int \eta(x', x)\, \varphi(x')\, dx' \tag{15.64}$$

where

$$\eta(x', x) = \int e^{i\lambda} u^*(\lambda, x') u(\lambda, x)\, d\lambda. \tag{15.65}$$

Note that the eigenfunctions depend on θ and τ. This method has the disadvantage that the eigenvalue problem must be solved.

For the discrete case we write

$$\{\theta A + \tau B\}\, u_n(x) = \lambda_n\, u_n(x) \tag{15.66}$$

and expand $\varphi(x)$ as

$$\varphi(x) = \sum_{n=0}^{\infty} c_n u_n(x) \tag{15.67}$$

with

$$c_n = \int u_n^*(x)\varphi(x)\, dx. \tag{15.68}$$

Therefore

$$e^{i\theta A+i\tau B}\varphi(x) = \sum_{n=0}^{\infty} e^{i\lambda_n} c_n u_n(x) \tag{15.69}$$

$$= \sum_{n=0}^{\infty} e^{i\lambda_n} \int u_n^*(x')\, \varphi(x')\, u_n(x)\, dx \tag{15.70}$$

and we can again write

$$e^{i\theta A + i\tau B}\varphi(x) = \int \eta(x', x)\,\varphi(x')\,dx' \tag{15.71}$$

where now

$$\eta(x', x) = \sum_{n=0}^{\infty} e^{i\lambda_n} u_n^*(x') u_n(x). \tag{15.72}$$

15.4.3 Linear Combination of X and D

We take for A and B a linear combination of the operators X and D:

$$A = c_{11}X + c_{12}D, \qquad B = c_{21}X + c_{22}D \tag{15.73}$$

which may be written in matrix form,

$$\begin{pmatrix} A \\ B \end{pmatrix} = \begin{pmatrix} c_{11} & c_{12} \\ c_{21} & c_{22} \end{pmatrix} \begin{pmatrix} X \\ D \end{pmatrix} = c \begin{pmatrix} X \\ D \end{pmatrix}. \tag{15.74}$$

The commutator works out to be

$$[A, B] = i(c_{11}c_{22} - c_{12}c_{21}) = i\det(c) \tag{15.75}$$

where $\det(c)$ is the determinant of c.

Now, using Eq. (15.53), we have

$$e^{i\theta A + i\tau B} = e^{i(\theta c_{11} + \tau c_{21})X + i(\theta c_{12} + \tau c_{22})D} \tag{15.76}$$

$$= e^{i(\theta c_{11} + \tau c_{21})(\theta c_{12} + \tau c_{22})/2} e^{i(\theta c_{11} + \tau c_{21})X} e^{i(\theta c_{12} + \tau c_{22})D}. \tag{15.77}$$

The association is hence given by

$$G(A,B) = \iint \hat{\sigma}(\theta, \tau) e^{i(\theta c_{11} + \tau c_{21})(\theta c_{12} + \tau c_{22})/2} e^{i(\theta c_{11} + \tau c_{21})X} e^{i(\theta c_{12} + \tau c_{22})D}\,d\theta\,d\tau \tag{15.78}$$

and the operation on a state function by

$$G(A,B)\varphi(x) = \iint \hat{\sigma}(\theta, \tau) e^{i(\theta c_{11} + \tau c_{21})(\theta c_{12} + \tau c_{22})/2} e^{i(\theta c_{11} + \tau c_{21})x}$$
$$\times \varphi(x + \theta c_{12} + \tau c_{22})\,d\theta\,d\tau. \tag{15.79}$$

For the characteristic function we have

$$M(\theta, \tau) = \int \varphi^*(x)\,e^{i\theta A + i\tau B}\varphi(x)\,dx \tag{15.80}$$

$$= \int \varphi^*(x)\,e^{i(\tau c_{21} + \theta c_{11})X + i(\theta c_{12} + \tau c_{22})D}\varphi(x)\,dx \tag{15.81}$$

15.4. The $e^{i\theta A + i\tau B}$ Correspondence

which simplifies to

$$M(\theta,\tau) = \int \varphi^*(x - \tfrac{1}{2}(\theta c_{12} + \tau c_{22})) \, e^{i(\theta c_{11} + \tau c_{21})x} \varphi(x + \tfrac{1}{2}(\theta c_{12} + \tau c_{22})) \, dx \quad (15.82)$$

and gives the distribution

$$P(\alpha,\beta) = \frac{1}{4\pi^2} \iiint \varphi^*(x - \tfrac{1}{2}(\theta c_{12} + \tau c_{22})) \, \varphi(x + \tfrac{1}{2}(\theta c_{12} + \tau c_{22}))$$
$$\times e^{i(\theta c_{11} + \tau c_{21})x} e^{-i\theta\alpha - i\tau\beta} \, d\theta d\tau dx \quad (15.83)$$

This distribution yields the following marginals,

$$\int P(\alpha,\beta) \, d\alpha = \left| \frac{1}{\sqrt{2\pi c_{22}}} \int \varphi(x) e^{-i(\beta x - c_{21} x^2/2)/c_{22}} dx \right|^2, \quad (15.84)$$

$$\int P(\alpha,\beta) \, d\beta = \left| \frac{1}{\sqrt{2\pi c_{12}}} \int \varphi(x) e^{-i(\alpha x - c_{11} x^2/2)/c_{12}} dx \right|^2. \quad (15.85)$$

Example. Rotation in Phase-Space. If we consider rotation in phase-space, then

$$c = \begin{pmatrix} \cos\phi & -\sin\phi \\ \sin\phi & \cos\phi \end{pmatrix} \quad (15.86)$$

which gives

$$M(\theta,\tau) = \int \varphi^*(x - \tfrac{1}{2}(-\theta \sin\phi + \tau \cos\phi)) \varphi(x + \tfrac{1}{2}(-\theta \sin\phi + \tau \cos\phi)) \quad (15.87)$$
$$\times e^{i(\theta \cos\phi + \tau \sin\phi)x} dx \quad (15.88)$$

for the characteristic function and the distribution is then

$$P(\alpha,\beta) = \frac{1}{4\pi^2} \iiint \varphi^*(x - \tfrac{1}{2}(-\theta \sin\phi + \tau \cos\phi)) \, \varphi(x + \tfrac{1}{2}(-\theta \sin\phi + \tau \cos\phi))$$
$$\times e^{i(\theta \cos\phi + \tau \sin\phi)x} e^{-i\theta\alpha - i\tau\beta} d\theta d\tau dx. \quad (15.89)$$

The marginals are,

$$\int P(\alpha,\beta) \, d\alpha = \left| \frac{1}{\sqrt{2\pi \cos\phi}} \int \varphi(x) e^{-i(\beta x - \sin\phi x^2/2)/\cos\phi} dx \right|^2, \quad (15.90)$$

$$\int P(\alpha,\beta) d\beta = \left| \frac{1}{\sqrt{2\pi \sin\phi}} \int \varphi(x) e^{i(\alpha x - \cos\phi x^2/2)/\sin\phi} dx \right|^2. \quad (15.91)$$

Example. Linear Motion. In quantum mechanics, D is the momentum operator when the Planck constant is taken to be one. For a free particle, the position operator changes according to

$$X(t) = X + Dt \quad (15.92)$$

where X is the position operator at time zero and where we have taken the mass to be 1. Hence it is of interest to consider

$$A = X + D t_1, \qquad B = X + D t_2 \qquad (15.93)$$

where A and B are the position operators at two different times, t_1 and t_2. The c matrix is now

$$c = \begin{pmatrix} 1 & t_1 \\ 1 & t_2 \end{pmatrix} \qquad (15.94)$$

with determinant

$$\det(c) = t_2 - t_1. \qquad (15.95)$$

The characteristic function is

$$M(\theta, \tau) = \int \varphi^*(x - \tfrac{1}{2}(\theta t_1 + \tau t_2)) \, e^{i(\theta+\tau)x} \varphi(x + \tfrac{1}{2}(\theta t_1 + \tau t_2)) \, dx \qquad (15.96)$$

which gives the distribution

$$P(\alpha, \beta) = \frac{1}{4\pi^2} \iiint \varphi^*(x - \tfrac{1}{2}(\theta t_1 + \tau t_2)) \, \varphi(x + \tfrac{1}{2}(\theta t_1 + \tau t_2))$$
$$\times e^{i(\theta+\tau)x} e^{-i\theta\alpha} e^{-i\tau\beta} d\theta d\tau dx. \qquad (15.97)$$

This distribution satisfies the marginals,

$$\int P(\alpha, \beta) \, d\alpha = \left| \frac{1}{\sqrt{2\pi t_2}} \int \varphi(x) e^{-i(\beta x - x^2/2)/t_2} dx \right|^2, \qquad (15.98)$$

$$\int P(\alpha, \beta) d\beta = \left| \frac{1}{\sqrt{2\pi t_1}} \int \varphi(x) e^{-i(\alpha x - x^2/2)/t_1} dx \right|^2. \qquad (15.99)$$

Example. X and $X + D$. We let

$$A = X, \qquad B = X + D. \qquad (15.100)$$

Although this is a special case of the previous example it is illustrative to do it from scratch by two different methods. The commutator is

$$[X, X + D] = [X, D] = i \qquad (15.101)$$

and using Eq. (15.53) we have

$$e^{i\theta X + i\tau(X+D)} = e^{i(\theta+\tau)(X+\tau/2)} e^{i\tau D} \qquad (15.102)$$

and further

$$e^{i\theta X + i\tau(X+D)} \varphi(x) = e^{i(\theta+\tau)(x+\tau/2)} \varphi(x+\tau). \qquad (15.103)$$

15.4. The $e^{i\theta A + i\tau B}$ Correspondence

Therefore
$$G(X, B) = \iint \hat{\sigma}(\theta, \tau) e^{i(\theta+\tau)(x+\tau/2)} e^{i\tau D} d\theta\, d\tau \qquad (15.104)$$

and for the operation on a function we have
$$G(X, B)\varphi(x) = \iint \hat{\sigma}(\theta, \tau) e^{i(\theta+\tau)(x+\tau/2)} \varphi(x+\tau)\, d\theta\, d\tau \qquad (15.105)$$
$$= \iint \hat{\sigma}(\theta, \tau - x) e^{i(\theta+\tau-x)(x+\tau)/2} \varphi(\tau)\, d\theta\, d\tau. \qquad (15.106)$$

Expressing this in terms of the symbol one obtains
$$G(X, B)\varphi(x) = \frac{1}{2\pi} \iint \sigma\left(\frac{x+\tau}{2}, \xi\right) e^{-i(\tau-x)[\xi - (\tau+x)/2]} \varphi(\tau)\, d\tau\, d\xi. \qquad (15.107)$$

We now obtain the same result by the method of Sec. 15.4.1 leading to Eq. (15.64). The eigenvalue problem is
$$\{\theta x + \tau(X + D)\} u(\lambda, x) = \lambda u(\lambda, x) \qquad (15.108)$$

or
$$\left\{(\theta + \tau)x - i\tau \frac{d}{dx}\right\} u(\lambda, x) = \lambda u(\lambda, x). \qquad (15.109)$$

The eigenfunctions normalized to a delta function are easily obtained
$$u(\lambda, x) = \frac{1}{\sqrt{2\pi\tau}} e^{i[\lambda x - (\theta+\tau)x^2/2]/\tau} \qquad (15.110)$$

and the calculation of $\eta(x', x)$ in Eq. (15.65) yields
$$\eta(x', x) = \delta(x' - \tau - x) e^{i(\theta+\tau)(\tau/2 + x)}. \qquad (15.111)$$

Therefore
$$e^{i\theta X + i\tau(X+D)} \varphi(x) = \int \eta(x', x) \varphi(x')\, dx' \qquad (15.112)$$
$$= \int \delta(x' - \tau - x) e^{i(\theta+\tau)[\tau/2 + x]} \varphi(x')\, dx' \qquad (15.113)$$
$$= e^{i(\theta+\tau)(\tau/2 + x)} \varphi(\tau + x) \qquad (15.114)$$

giving
$$G(X, B) = \iint \hat{\sigma}(\theta, \tau) e^{i(\theta+\tau)(x+\tau/2)} e^{i\tau D} d\theta\, d\tau \qquad (15.115)$$

which is the same as Eq. (15.104).

For the characteristic function we have
$$M(\theta, \tau) = \int \varphi^*(x - \tfrac{1}{2}\tau)) e^{i(\theta+\tau)x} \varphi(x + \tfrac{1}{2}\tau)\, dx \qquad (15.116)$$

which leads to the distribution

$$P(\alpha, \beta) = \frac{1}{2\pi} \int \varphi^*(\alpha - \tfrac{1}{2}\tau) \, e^{-i(\beta-\alpha)\tau} \varphi(\alpha + \tfrac{1}{2}\tau) \, d\tau \qquad (15.117)$$

which can be written as

$$P(\alpha, \beta) = W(\alpha, \beta - \alpha) \qquad (15.118)$$

where $W(\alpha, \beta)$ is the Wigner distribution of α and β.

Example. X **and** D^2. Consider the case where

$$A = X, \qquad B = D^2. \qquad (15.119)$$

This does not fit into our special cases discussed in Sec. 15.4.1 since

$$[X, D^2] = 2Di \qquad (15.120)$$

but we can do it using the general theorem given in Sec. 15.4.1. We solve the eigenvalue problem

$$\{ \theta X + \tau D^2 \} \, \widehat{u}(\lambda, p) = \lambda \, \widehat{u}(\lambda, p) \qquad (15.121)$$

in the Fourier domain

$$\{ i\theta \frac{d}{dp} + \tau p^2 \} \widehat{u}(\lambda, p) = \lambda \, \widehat{u}(\lambda, p). \qquad (15.122)$$

The eigenfunctions are

$$\widehat{u}(\lambda, p) = \frac{1}{\sqrt{2\pi\theta}} e^{-i(\lambda p - \tau p^3/3)/\theta} \qquad (15.123)$$

and using Eq. (15.65) we have

$$\eta(p, p'; \theta, \tau) = \delta(p' - p + \theta) e^{-i\tau(p^3 - p'^3)/3\theta}. \qquad (15.124)$$

According to Eq. (15.64), we then have

$$e^{i(\theta X + \tau D^2)} \widehat{\varphi}(p) = \int \eta(p, p'; \theta, \tau) \, \widehat{\varphi}(p') \, dp' \qquad (15.125)$$

$$= \int \delta(p' - p + \theta) e^{i\tau(p^3 - p'^3)/3\theta} \, \widehat{\varphi}(p') \, dp' \qquad (15.126)$$

$$= e^{i\tau(p^2 - \theta p + \theta^2/3)} \, \widehat{\varphi}(p - \theta) \qquad (15.127)$$

$$= e^{i\tau(p^2 - \theta p + \theta^2/3)} \, e^{i\theta X} \widehat{\varphi}(p) \qquad (15.128)$$

and therefore

$$G(X,B)\widehat{\varphi}(p) = \iint \widehat{\sigma}(\theta, \tau) e^{i\tau(p^2 - \theta p + \theta^2/3)} \, \widehat{\varphi}(p - \theta) \, d\theta \, d\tau \qquad (15.129)$$

15.4. The $e^{i\theta A + i\tau B}$ Correspondence

and
$$G(X,B) = \iint \hat{\sigma}(\theta,\tau) e^{i\tau(p^2 - \theta p + \theta^2/3)} e^{i\theta X} \, d\theta \, d\tau. \tag{15.130}$$

Evaluating the characteristic function, Eq. (15.45), one obtains
$$M(\theta,\tau) = \int \hat{\varphi}^*(p) e^{i\tau(p^2 - \theta p + \theta^2/3)} \hat{\varphi}(p-\theta) \, dp \tag{15.131}$$
$$= \int \hat{\varphi}^*(p + \tfrac{1}{2}\theta) e^{i\tau(p^2 + \theta^2/12)} \hat{\varphi}(p - \tfrac{1}{2}\theta) \, dp \tag{15.132}$$

and the distribution works out to be
$$P(\alpha,\beta) = \frac{1}{2\pi} \iint \hat{\varphi}^*(p + \tfrac{1}{2}\theta)\hat{\varphi}(p - \tfrac{1}{2}\theta)\delta(p^2 + \theta^2/12 - \beta)e^{-i\theta\alpha} \, dp \, d\theta. \tag{15.133}$$

Furthermore, for the expectation value of a symbol, we have
$$\langle \sigma(\alpha,\beta) \rangle = \iint \sigma(\alpha,\beta) P(\alpha,\beta) \, d\alpha \, d\beta \tag{15.134}$$
$$= \frac{1}{2\pi} \iiint \sigma(\alpha, p^2 + \tfrac{1}{12}\theta^2) \hat{\varphi}^*(p + \tfrac{1}{2}\theta)\hat{\varphi}(p - \tfrac{1}{2}\theta) e^{-i\theta\alpha} \, dp \, d\theta \, d\alpha. \tag{15.135}$$

Example. X and Scale. The scale operator, C, is
$$C = \tfrac{1}{2}(xD + Dx) = \frac{1}{2i}\left(x\frac{d}{dx} + \frac{d}{dx}x\right) \tag{15.136}$$

in the spatial domain, and has the property
$$e^{i\tau C}\varphi(x) = e^{\tau/2}\varphi(e^\tau x). \tag{15.137}$$

The commutation is readily evaluated,
$$[X,C] = iX \tag{15.138}$$

and this fits the case given by Eq. (15.57) with $\xi = i$. We therefore have
$$e^{i\theta X + i\tau C} = e^{i\theta\mu X} e^{i\tau C} e^{i\theta X} \tag{15.139}$$

with
$$\mu = \frac{(1-\tau)e^\tau - 1}{\tau}. \tag{15.140}$$

Now consider
$$e^{i\tau C} e^{i\theta X}\varphi(x) = e^{\tau/2} e^{i\theta e^\tau x}\varphi(e^\tau x) = e^{i\theta e^\tau x} e^{\tau/2}\varphi(e^\tau x) = e^{i\theta e^\tau x} e^{i\tau C}\varphi(x) \tag{15.141}$$

and therefore we have
$$e^{i\tau C} e^{i\theta X} = e^{i\theta e^\tau x} e^{i\tau C}. \tag{15.142}$$

Substituting into Eq. (15.139), we have
$$e^{i\theta X + i\tau C} = e^{i\theta(\mu + e^\tau)X} e^{i\tau C}. \tag{15.143}$$

The operator corresponding to the symbol is
$$G(X,C) = \iint \hat{\sigma}(\theta, \tau) e^{i\theta(\mu + e^\tau)X} e^{i\tau C} d\theta d\tau \tag{15.144}$$

and further for the operation on a function,
$$G(X,C)\varphi(x) = \iint \hat{\sigma}(\theta, \tau) e^{i\theta(\mu + e^\tau)X} e^{\tau/2} \varphi(e^\tau x) d\theta d\tau. \tag{15.145}$$

Using the fact
$$(\mu + e^\tau) = (e^\tau - 1)/\tau = \frac{2}{\tau} e^{\tau/2} \sinh(\tau/2) \tag{15.146}$$

Eq. (15.145) simplifies to
$$G(X,C)\varphi(x) = \iint \hat{\sigma}(\theta, \tau) e^{i[2\theta x e^{\tau/2} \sinh(\tau/2)]/\tau} e^{\tau/2} \varphi(e^\tau x) d\theta d\tau. \tag{15.147}$$

The characteristic function is
$$M(\theta, \tau) = \int \varphi^*(x) e^{i\theta X + i\tau C} \varphi(x) dx \tag{15.148}$$

which, after using the above relations, gives
$$M(\theta, \tau) = \int \varphi^*(e^{-\tau/2} x) e^{2i\theta x \sinh(\sigma/2)/\tau} \varphi(e^{\tau/2} x) dx \tag{15.149}$$

and for the distribution, one obtains
$$P(x, c) = \frac{1}{4\pi^2} \int \varphi^*(e^{-\tau/2} x') e^{2i\theta x' \sinh(\tau/2)/\tau} \varphi(e^{\sigma/2} x') e^{-i\theta x - i\tau c} d\theta d\tau dx' \tag{15.150}$$

$$= \frac{1}{2\pi} \int \frac{\tau e^{-i\tau c}}{2 \sinh \tau/2} \varphi^*\left(e^{-\tau/2} \frac{\tau x}{2 \sinh \tau/2}\right) \varphi\left(e^{\tau/2} \frac{\tau x}{2 \sinh \tau/2}\right) d\tau. \tag{15.151}$$

Example. X and 1/D. We now consider the operators X and $1/D$ where $1/D$ can be thought of as the inverse frequency. We define
$$A = X \qquad B = R = p_0/D \tag{15.152}$$

15.4. The $e^{i\theta A + i\tau B}$ Correspondence

where p_0 is a scaling factor. The commutator of X and R works out to be

$$[X, R] = -\frac{i}{p_0} R^2. \tag{15.153}$$

To evaluate $e^{i\theta X + i\tau R}$, we use the method described in Sec. 15.4.1, where we have to solve the eigenvalue problem

$$(\theta X + \tau R)\widehat{u}(\lambda, p) = \lambda \widehat{u}(\lambda, p). \tag{15.154}$$

Explicitly

$$\left[i\theta \frac{d}{dp} + \tau \frac{p_0}{p}\right] \widehat{u}(\lambda, p) = \lambda \widehat{u}(\lambda, p) \tag{15.155}$$

and the solution is

$$\widehat{u}(\lambda, p) = \frac{1}{\sqrt{2\pi\theta}} e^{-i(\lambda p - p_0 \tau \ln |p|)/\theta}. \tag{15.156}$$

Substituting into Eq. (15.65), one obtains

$$\eta(p, p') = e^{ip_0 \tau \ln(p/p')/\theta} \delta(\theta + p' - p) \tag{15.157}$$

which gives

$$e^{i\theta X + i\tau R} \widehat{\varphi}(p) = \widehat{\varphi}(p - \theta) \exp\left[i\frac{p_0 \tau}{\theta} \ln \frac{p}{p - \theta}\right]. \tag{15.158}$$

Hence

$$G(X, R)\widehat{\varphi}(p) = \iint \widehat{\sigma}(\theta, \tau) \exp\left[i\frac{p_0 \tau}{\theta} \ln \frac{p}{p - \theta}\right] \widehat{\varphi}(p - \theta) \, d\theta \, d\tau. \tag{15.159}$$

The characteristic function is

$$M(\theta, \tau) = \int \widehat{\varphi}^*(p) e^{i\theta X + i\tau R} \widehat{\varphi}(p) \, dp \tag{15.160}$$

which evaluates to

$$M(\theta, \tau) = \int \widehat{\varphi}^*(p + \tfrac{1}{2}\theta) \exp\left[i\frac{p_0 \tau}{\theta} \ln \frac{p + \tfrac{1}{2}\theta}{p - \tfrac{1}{2}\theta}\right] \widehat{\varphi}(p - \tfrac{1}{2}\theta) \, dp. \tag{15.161}$$

The distribution is given by

$$\Gamma(x, r) = \frac{1}{4\pi^2} \int \widehat{\varphi}^*(p + \tfrac{1}{2}\theta) \exp\left[i\frac{p_0 \tau}{\theta} \ln \frac{p + \tfrac{1}{2}\theta}{p - \tfrac{1}{2}\theta}\right] \widehat{\varphi}(p - \tfrac{1}{2}\theta) e^{-i\theta x - i\tau r} \, d\theta \, d\tau \, dp \tag{15.162}$$

which simplifies to [19, 70, 73, 8]

$$P(x, r) = \frac{1}{8\pi r^3} \int \left(\frac{p_0 y}{\sinh y/2}\right)^2 e^{-ip_0 y x/r} \widehat{\varphi}^*\left(\frac{p_0}{2r} \frac{y e^{y/2}}{\sinh y/2}\right) \widehat{\varphi}\left(\frac{p_0}{2r} \frac{y e^{-y/2}}{\sinh y/2}\right) dy. \tag{15.163}$$

15.5 The $e^{i\theta A} e^{i\tau B}$ Correspondence

For the operator association, we take

$$G(A,B) = \iint \hat{\sigma}(\theta, \tau) e^{i\theta A} e^{i\tau B} d\theta\, d\tau \qquad (15.164)$$

and for the operation on a function, we write

$$G(A,B)\varphi(x) = \iint \hat{\sigma}(\theta, \tau) e^{i\theta A} e^{i\tau B} \varphi(x)\, d\theta\, d\tau. \qquad (15.165)$$

Often the operation of $e^{i\theta A} e^{i\tau B} \varphi(x)$ can be carried out explicitly as in some of the examples we consider. However, a general result can be obtained in terms of the eigenfunctions of A and B. Set

$$A\, v(a,x) = a\, v(a,x), \qquad (15.166)$$
$$B\, u(b,x) = b\, u(b,x), \qquad (15.167)$$

and suppose that the eigenfunctions are connected by

$$u(b,x) = \int T(a,b) v(a,x)\, da \qquad (15.168)$$

where $T(a,b)$ is called the transformation matrix and is given by

$$T(a,b) = \int v^*(a,x) u(b,x)\, dx. \qquad (15.169)$$

Also,

$$v(a,x) = \int T^*(a,b) u(b,x)\, db. \qquad (15.170)$$

The state function can be expanded in either representation

$$\varphi(x) = \int v(a,x) F_a(a)\, da \qquad (15.171)$$

or

$$\varphi(x) = \int u(b,x) F_b(b)\, db \qquad (15.172)$$

with

$$F_a(a) = \int v^*(a,x)\, \varphi(x)\, dx, \qquad (15.173)$$

$$F_b(b) = \int u^*(b,x)\, \varphi(x)\, dx. \qquad (15.174)$$

15.5. The $e^{i\theta A}e^{i\tau B}$ Correspondence

Therefore

$$e^{i\theta A} e^{i\tau B} \varphi(x) = e^{i\theta A} \int e^{i\tau b} u(b,x) F_b(b)\, db \qquad (15.175)$$

$$= e^{i\theta A} \iint e^{i\tau b} T(a,b) v(a,x) F_b(b)\, da\, db \qquad (15.176)$$

yielding that

$$e^{i\theta A} e^{i\tau B} \varphi(x) = \iint e^{i\theta a} e^{i\tau b} T(a,b) v(a,x) F_b(b)\, da\, db. \qquad (15.177)$$

Hence, for the characteristic function we have

$$M(\theta,\tau) = \int \varphi^*(x) e^{i\theta A} e^{i\tau B} \varphi(x)\, dx \qquad (15.178)$$

$$= \iiint \varphi^*(x) e^{i\theta a} e^{i\tau b} T(a,b) v(a,x) F_b(b)\, da\, db\, dx \qquad (15.179)$$

which simplifies to

$$M(\theta,\tau) = \iint F_a^*(a)\, e^{i\theta a} e^{i\tau b} T(a,b) F_b(b)\, da\, db. \qquad (15.180)$$

The distribution is then

$$P(a,b) = F_a^*(a)\, T(a,b) F_b(b). \qquad (15.181)$$

Example. (X, D) **case.** We take

$$A = X, \quad B = D. \qquad (15.182)$$

The eigenvalue problems

$$X\, v(a,x) = a\, v(a,x), \qquad (15.183)$$
$$D\, u(b,x) = b\, u(b,x) \qquad (15.184)$$

give

$$v(a,x) = \delta(x-a), \qquad (15.185)$$
$$u(b,x) = \frac{1}{\sqrt{2\pi}} e^{ibx}, \qquad (15.186)$$

and for the transformation matrix, Eq. (15.169), we have

$$T(a,b) = \int v^*(a,x) u(b,x)\, dx = \frac{1}{\sqrt{2\pi}} e^{iab}. \qquad (15.187)$$

Hence, for a state function $\varphi(x)$, the transforms are

$$F_a(a) = \int \delta(x-a)\,\varphi(x)\,dx = \varphi(a) \tag{15.188}$$

and

$$F_b(b) = \frac{1}{\sqrt{2\pi}} \int e^{-ibx}\,\varphi(x)\,dx = \widehat{\varphi}(b) \tag{15.189}$$

and therefore the distribution is given by Eq. (15.181)

$$P(a,b) = F_a^*(a)\,T(a,b)F_b(b) \tag{15.190}$$
$$= \frac{1}{\sqrt{2\pi}}\varphi(a)\,e^{iba}\widehat{\varphi}(b) \tag{15.191}$$

which, as expected, is the Margenau-Hill distribution with $a = x$ and $b = p$.

Example: Scale and X. We take

$$A = C = \frac{1}{2}(XD+DX), \qquad B = X. \tag{15.192}$$

The association is

$$G(C,X) = \iint \widehat{\sigma}(\theta,\tau)e^{i\tau C}\,e^{i\theta X}\,d\theta\,d\tau \tag{15.193}$$

and using Eq. (15.137) we obtain

$$G(C,X)\,\varphi(x) = \iint \widehat{\sigma}(\theta,\tau)\,e^{i\tau C}\,e^{i\theta x}\varphi(x)\,d\theta\,d\tau \tag{15.194}$$
$$= \iint \widehat{\sigma}(\theta,\tau)\,e^{\tau/2}\,e^{i\theta e^\tau x}\varphi(e^\tau x)\,d\theta\,d\tau \tag{15.195}$$

which simplifies to

$$G(C,X)\,\varphi(x) = \frac{1}{2\pi}\iint \sigma(e^\tau x,\xi)\,e^{-i\tau\xi}\,e^{\tau/2}\,\varphi(e^\tau x)d\xi\,d\tau. \tag{15.196}$$

15.6 The $e^{i\tau A/2}e^{i\theta B}e^{i\tau A/2}$ Correspondence

For this correspondence, we just take the example of position and scale [1, 19, 53]

$$A = X, \qquad B = C = \tfrac{1}{2}(XD+DX). \tag{15.197}$$

Hence

$$G(X,C)\,\varphi(x) = \iint \widehat{\sigma}(\theta,\tau)e^{i\tau C/2}e^{i\theta X}e^{i\tau C/2}\,\varphi(x)\,d\theta\,d\tau. \tag{15.198}$$

15.6. The $e^{i\tau A/2}e^{i\theta B}e^{i\tau A/2}$ Correspondence

Using Eq. (15.137) we obtain that

$$e^{i\tau C/2}e^{i\theta X}e^{i\tau C/2}\varphi(x) = e^{i\theta e^{\tau/2}x}e^{\tau/2}\varphi(e^{\tau}x) \qquad (15.199)$$

and the operation on a function is

$$G(X,C)\varphi(x) = \iint \hat{\sigma}(\theta,\tau)\, e^{i\theta e^{\tau/2}x}\, e^{\tau/2}\, \varphi(e^{\tau}x)\, d\theta\, d\tau. \qquad (15.200)$$

For the characteristic function we have

$$M(\theta,\tau) = \int \varphi(x)\, e^{i\tau C/2}\, e^{i\theta X}\, e^{i\tau C/2}\, \varphi(x)\, dx \qquad (15.201)$$

$$= \int \varphi(x)\, e^{i\theta e^{\tau/2}x}\, e^{\tau/2}\, \varphi(e^{\tau}x)\, dx \qquad (15.202)$$

which simplifies to

$$M(\theta,\tau) = \int \varphi^*(e^{-\tau/2}x)\, e^{i\theta x}\, \varphi(e^{\tau/2}x)\, dx. \qquad (15.203)$$

The distribution evaluates to

$$P(x,c) = \frac{1}{2\pi}\int \varphi^*(e^{-\tau/2}x)\, e^{-i\tau c}\, \varphi(e^{\tau/2}x)\, d\tau. \qquad (15.204)$$

Bibliography

[1] R. A. Altes, Wideband, proportional–bandwidth Wigner–Ville analysis, IEEE Trans. Acoust., Speech, Signal Processing, 38, 1005–1012, 1990.

[2] M. G. Amin, Time–varying spectrum estimation of a general class of nonstationary processes, Proceedings of the IEEE, 74, 1800–1802, 1986.

[3] J. C. Andrieux, M. R. Feix, G. Mourgues, P. Bertrand, B. Izrar, and V. T. Nguyen, Optimum smoothing of the Wigner–Ville distribution, IEEE Trans. Acoust., Speech, Signal Processing, 35, 764–769, 1987.

[4] P. Boggiatto, G. De Donno, and A. Oliaro, Time-Frequency Representations of Wigner Type and Pseudo-differential Operators, Trans. Amer. Math. Soc. 362, 4955-498, 2010.

[5] R. G. Baraniuk and D. L. Jones, A signal–dependent time–frequency representation: optimal kernel design, IEEE Trans. on Signal Processing, 41, 1589–1602, 1993.

[6] R. G. Baraniuk and L. Cohen, On Joint Distributions for Arbitrary Variables, IEEE Signal Processing Letters, IEEE Signal Processing Letters, 2, 10-12 , 1995.

[7] R. G. Baraniuk and D. L. Jones, Unitary equivalence: a new twist on signal processing, IEEE Trans. on Signal Processing, 43, 2269-2282, 1995.

[8] P. Bertrand and J. Bertrand, Time–Frequency Representations of Broad Band Signals, in: The Physics of Phase Space, edited by Y. S. Kim and W. W. Zachary, Springer Verlag, 250–252, 1987.

[9] F. Bopp, La Mechanique Quantique Est–Elle Une Mechanique Statistique Classique Particuliere?, Ann. L'lust. II. Poincare, 15, 81–112, 1956.

[10] M. Born and P. Jordan, Zur Quantenmechanik, Zeit. f. Phys., 34, 858–888, 1925.

[11] H. I. Choi and W. J. Williams, Improved time–frequency representation of multicomponent signals using exponential kernels, IEEE Trans. on Acoust., Speech, Signal Processing, 37, 862–871, 1989.

[12] T.A.C.M Claasen and W.F.G. Mecklenbrauker, The Wigner distribution-A tool for time-frequency signal analysis - Part III: Relations with other time-frequency signal transformations, Philips Journal of Research, 35, 372-389, 1980.

[13] L. Cohen, Generalized phase–space distribution functions, Jour. Math. Phys., 7, 781–786, 1966.

[14] L. Cohen, Expansion Theorem for Functions of Operators, Jour. Math. Phys. 7, 244, 1966.

[15] Hamiltonian Operator via Feynman Path Integral, Jour. Math. Phys. 11, 3296-3297, 1970; Correspondence Rules and Path Integrals, Jour. Math. Phys. 17, 597-598, 1976.

[16] L. Cohen, Quantization Problem and Variational Principle in the Phase Space Formulation of Quantum Mechanics, J. Math. Phys., 17, 1863, 1976.

[17] L. Cohen, Local Kinetic Energy in Quantum Mechanics, J. Chem Phys. 70, 788, 1979.

[18] L. Cohen, Time-Frequency Distributions - A Review, Proc. of the IEEE, 77, 941-981, 1989.

[19] L. Cohen, The scale representation, IEEE Trans. on Signal Processing, 41, 3275–3293, 1993.

[20] L. Cohen, *Time-Frequency Analysis*, Prentice-Hall, Englewood Cliffs, 1995.

[21] L. Cohen, A General Approach for Obtaining Joint Representation in Signal Analysis. Part I. Characteristic Function Operator Method, IEEE Transactions on Signal Processing, 44, 1080-1090, 1996; Part II. General Class, Mean and Local Values, and Bandwidths, IEEE Transactions on Signal Processing, 44, 1091-1098, 1996.

[22] L. Cohen, Wigner quasi-probability distributions for arbitrary operators, J. Mod. Optics, 51, 2761-2769, 2004.

[23] L. Cohen, The Weyl Transform and its Generalization, Rend. Sem. Mat. Univ. Pol. Torino - Vol. 66, 4, 1-12, 2008.

[24] G. Cristobal, C. Gonzalo and J. Bescos, Image Filtering and Analysis through the Wigner Distribution, in: Advances in Electronics and Electron Physics, P. E. Hawkes, ed, Academic Press, 1990.

[25] K. Davidson, Instantaneous moments of signals, Ph.D. dissertation, University of Pittsburgh, 2001.

[26] N. O. D. Bruijn, A theory of generalized functions, with applications to Wigner distribution and Weyl correspondence, Nieuw Arch. Wisk. (4). 3. no. 21, 205-280. 1973.

[27] J. S. Dowker, Path integrals and ordering rules, J. Math. Phys. 17, 1873-1874, 1976.

[28] P. D. Finch and R. Groblicki, Bivariate probability densities with given margins, Found. of Phys., 14, 549–552, 1984.

[29] L. Galleani and L. Cohen, The Wigner distribution for classical systems, Physics Letters A, 302, 149-155, 2002.

[30] L. Galleani, L. Cohen, Time Frequency Wigner Distribution Approach To Differential Equations, in: *Nonlinear Signal and Image Processing: Theory, Methods, and Applications*, K. Barner and G. Arce (Eds.), CRC Press, 2003.

[31] L. Hormander, The Weyl calculus of pseudodifferential operators, Comm. Pure Appl. Math., voL 32, 359-443, 1979.

[32] D. Gabor, Theory of communication, IEE J. Comm. Engr., 93, 429-457, 1946.

[33] H. J. Groenewold, On the Principles of elementary quantum mechanics, Physica, 12, 405, 1946.

[34] K. Husimi, Some Formal Properties of the Density Matrix, Proc. Phys. Math. Soc. Japan, 22, 264-314, 1940.

[35] L. D. Jacobson and H. Wechsler, Joint spatial/spatial–frequency representations, Signal Processing, 14, 37–68, 1988.

[36] M. Jammer, *The Philosophy of Quantum Mechanics*, John Wiley & Sons, 1974.

[37] A. J. E. M. Janssen, On the locus and spread of pseudo–density functions in the time–frequency plane, Philips Journal of Research, 37, 79–110, 1982.

[38] A. J. E. M. Janssen, Bilinear phase–plane distribution functions and positivity, Jour. Math. Phys., 26, 1986–1994, 1985.

[39] J. Jeong and W. Williams, Kernel design for reduced interference distributions, IEEE Trans. Sig. Proc., 40, 402–412, 1992.

[40] E. H. Kerner and W. G. Sutcliffe, Unique Hamiltonian Operators via Feynman Path Integrals, J. Math. Phys. 11, 391-393, 1970.

[41] A. Khinchin, Bull. Univ. Moscow, 1, 1937.

[42] M. Kim and M. O. Scully, private communication, 2009.

[43] J. G. Kirkwood, Quantum statistics of almost classical ensembles, Phys. Rev., 44, 31–37, 1933.

[44] V. Kuryshkin, Some problems of quantum mechanics possessing a non–negative phase–space distribution functions, International Journal of Theoretical Physics, 7, 451–466, 1973.

[45] V. Kuryshkin V., Lybas I. and Zaparovanny Yu., Sur le problem de la regle de correspondence en theorie quantique, Ann. Fond. L. de Broglie 3, 45-61, 1978.

[46] E. Lukacs, *Characteristic Functions*, Charles Griffin and Company, London, 1970

[47] B. Leaf, Weyl Transformation and the Classical Limit of Quantum Mechanics, J. Math. Phys. 9, 65-72, 1968.

[48] H. W. Lee, Theory and application of the quantum phase-space distribution-functions, Physics Reports, 259, 147-211, 1995.

[49] P. Loughlin, J. Pitton and L. Atlas, Bilinear time-frequency representations: new insights and properties, IEEE Trans. Sig. Process., 41, 750-767, 1993.

[50] P. Loughlin and K. Davidson, Instantaneous kurtosis, IEEE Signal Process. Lets., 7, 156-159, 2000.

[51] H. Margenau and R. N. Hill, Correlation between measurements in quantum theory, Prog. Theoret. Phys., 26, 722–738, 1961.

[52] H. Margenau and L. Cohen, Probabilities in quantum mechanics, in: *Quantum Theory and Reality*, edited by M. Bunge, Springer–Verlag, 1967.

[53] N. M. Marinovich, *The Wigner distribution and the ambiguity function: Generalizations, enhancement, compression and some applications*, Ph. D. Dissertation, The City University of New York, NY, 1986.

[54] I. W. Mayes and J. S. Dowker, Hamiltonian orderings and functional Integrals, J. Math. Phys. 14, 434-439, 1973.

[55] N. H. McCoy, On the Function in Quantum Mechanics Which Corresponds to a Given Function in Classical Mechanics, Proc. Natl. Acad. Sci. U.S.A., 18, 674–676, 1932.

[56] C. L. Mehta, Phase-Space Formulation of the Dynamics of Canonical Variables, J. Math. Phys., 5, 677-686, 1964.

[57] M. M. Mizrahi, The Weyl correspondence and path integrals, J. Math. Phys. 16, 2201-2206, 1975.

[58] J. M. Morris and D-S. Wu, On Alias–Free Formulations of Discrete–Time Cohen's Class of Distributions, IEEE Trans. on Signal Processing, Vol. 44, 1355 - 1364 , 1996.

[59] G. Mourgues, M. R. Feix, J. C. Andrieux and P. Bertrand, Not necessary but sufficient conditions for the positivity of generalized Wigner functions, Jour. Math. Phys., 26, 2554–2555, 1985.

[60] J. E. Moyal, Quantum mechanics as a statistical theory, Proc. Camb. Phil. Soc., 45, 99–124, 1949.

[61] J. G. Muga, J. P. Palao and R. Sala Average local values and local variances in quantum mechanics, Phys. Lett. A 238, 90-94, 1998.

[62] J. G. Muga, D. Seidel, and G. C. Hegerfeldt, Quantum kinetic energy densities: an operational approach, J. of Chem. Phys., 122, 154016-1, 2005.

[63] M. Mugur–Schachter, A study of Wigner's theorem on joint probabilities, Found. of Phys., 9, 389–404, 1979.

[64] R. F. O'Connell, The Wigner distribution function – 50th birthday, Found. of Phys., 13, 83–92, 1983.

[65] S. Oh and R. J. Marks II, Some properties of the generalized time–frequency representation with cone shaped kernel, IEEE Trans. Sig. Proc., 40, 1735–1745, 1992.

[66] A. Perez, *Quantum Theory: Concepts and Methods*, Kluwar, 1995.

[67] J. Pitton, The Statistics of Time-Frequency Analysis, J. Franklin Institute, 337, 379-388, 2000.

[68] M. Poletti, The Homomorphic Analytic Signal, IEEE Trans. Signal Processing, 45, 1943-1953, 1997.

[69] J. C. T. Pool, Mathematical aspects of the Weyl correspondence, J. Math. Phys. 7, 66-76, 1966.

[70] T. E. Posch, Wavelet Transform and Time–Frequency Distributions, Proc. SPIE, 1152, 477–482, 1988.

[71] T. Reed and H. Wechsler, Tracking on nonstationarities for texture fields, Signal Processing, 14, 95–102, 1988.

[72] A. W. Rihaczek, Signal energy distribution in time and frequency, IEEE Trans. Info. Theory, 14, 369–374, 1968.

[73] O. Rioul and P. Flandrin, Time–scale energy distributions: a general class extending wavelet transforms, IEEE Trans. on Signal Processing, 40, 1746–1757, 1992.

[74] G. J. Ruggeri, On phase space description of quantum mechanics, Prog. Theoret. Phys. 46, 1703–1712, 1971.

[75] R. Sala, J. P. Palao and J. G. Muga, Phase space formalisms of quantum mechanics with singular kernel, Phys. Lett. A 231, 304-310, 1997.

[76] R. Sala, J. G. Muga, Quantum methods for classical dynamics in Louiville space, Physics Letters, A231, 180-184, 1994.

[77] Sayeed, A.M. and D.L. Jones, Optimal Kernels for Non-Stationary Spectral Estimation, IEEE Trans. on Signal Processing, 43, 478–491, 1995.

[78] W.P. Schleich, *Quantum Optics in Phase Space*, Wiley, 2001.

[79] M. O. Scully and M. S. Zubairy, *Quantum Optics*, Cambridge University Press, Cambridge, England: 1997.

[80] M. O. Scully and L. Cohen, Quasi-Probability Distributions for Arbitrary Operators, in: *The Physics of Phase Space*, edited by Y.S. Kim and W.W. Zachary, Springer Verlag, New York, 1987.

[81] R. G. Shenoy and T. W. Parks, The Weyl Correspondence and time–frequency analysis, IEEE Trans. on Signal Processing, 42, 318–331, 1994.

[82] J. R. Shewell, On the formation of quantum-mechanical operatorsAm. J. Phys. 27, 16-21,1959.

[83] M. D. Srinivas and E. Wolf, Some nonclassical features of phase–space representations of quantum mechanics, Phys. Rev. D, 11, 1477–1485, 1975.

[84] R. Sutherland, Joint distribution indicated by the wave equations of quantum mechanics, J. Math. Phys. 23, 2389-2392, 1982.

[85] F. J. Testa, Quantum Operator Ordering and the Feynman Formulation, J. Math. Phys. 12, 1471, 1971.

[86] J. Tekel, private communication, 2012.

[87] A. Torre, *Linear Ray and Wave Optics in Phase Space*, Elsevier B. V., The Netherlands, 2005

[88] J. Ville, Theorie et applications de la notion de signal analytique, Cables et Transmissions, 2A, 61–74, 1948.

[89] H. Weyl, *The Theory of Groups and Quantum Mechanics*, E. P. Dutton and Co., 1928.

[90] B. White, Transition Kernels for Bilinear Time–Frequency Signal Representations, IEEE Trans. Signal Processing, 39, 542–544, 1991.

[91] M. W. Wong, *Weyl Transforms*, Springer-Verlag, New York, 1998.

[92] Yu. I. Zaparovanny, The Correspondence Principle between Classical and quantum quantities, Sov. Phys., 762-766, 1975.

[93] E. P. Wigner, On the quantum correction for thermodynamic equilibrium, Physical Review, 40, 749–759, 1932.

[94] Y. Zhao, L. E. Atlas, and R. J. Marks II, The use of cone–shaped kernels for generalized time–frequency representations of nonstationary signals, IEEE Trans. Acoust., Speech, Signal Processing, 38, 1084–1091, 1990.

[95] R. M. Wilcox, Exponential Operators and Parameter Differentiation in Quantum Physics, J. Math. Phys., 8, 962-982, 1967.

Index

adjoint, 6, 86
    of $e^{i\theta X+i\tau D}$, 11
    of translation operator, 9
amplitude of a signal, 100
anti-commutator, 4, 46, 55
anti-normal ordering, 82
anti-standard ordering, 72
arbitrary operators
    $e^{i\tau A/2}e^{i\theta B}e^{i\tau A/2}$ correspondence, 148
    $e^{i\theta A+i\tau B}$ correspondence, 135
    Baker-Cambell-Hausdorf formula, 10, 136
    commuting operators, 133
    disentanglement of $e^{i\theta A+i\tau B}$, 136
    linear combination of X and D, 138
    rotation in phase-space, 139
arbitrary operators: single operator, 111–119
arbitrary operators: two operators, 131–150
association rules, 3

Baker-Cambell-Hausdorf formula, 10
Born interpretation, 114
Born-Jordan ordering, 2, 58, 73, 78

c-function, 3
characteristic function approach, 63, 135
characteristic function operator, 10
Choi-Williams ordering, 74
classical function, 3

commutator, 1, 4, 7, 10, 11, 23, 45, 46, 55, 95, 121, 126, 131, 138, 140, 145
commuting operators, 133
complex signals, 98
correspondence rules, 1, 3

delta function, 17
delta function association, 56, 57
differential equations
    ordinary, 105
    partial, 106

eigenvalue problem in phase-space, 107
expectation value, 114

Fourier association, 56, 57
Fourier transform pairs, 4
functions of operators, 7, 16

Gabor, 99
Gaussian window, 80
generalized association, 47–59
    algebra, 52
    delta function association, 56, 57
    form of, 58
    Fourier association, 56, 57
    Hermitian adjoint, 53
    linearity, 52
    monomial from the kernel, 51
    monomial rule, 50
    operation on a function, 49
    operational form, 48
    operator to symbol, 50
    product of operators, 54

Taylor series association, 56
transformation between associations, 55
translation invariance, 53
unit correspondence, 52
generalized characteristic function operator, 12
generalized distribution, 2, 61–67, 70
relation between distributions, 64
transformation of, 88

Hermitian adjoint, 53
Hermitian operator, 86
Hilbert transform, 99

instantaneous frequency, 98, 100
inverse frequency, 118

kernel for
anti standard ordering, 72
anti-normal ordering, 82
Born-Jordan ordering, 73
Choi-Williams ordering, 74
normal ordering, 82
spectrogram, 78
standard ordering, 70
symmetrization ordering, 73
Weyl ordering, 75
ZAM ordering, 76
Khintchine theorem, 129–130

linear combination of X and D, 114

manipulating $D^m X^n$ and $X^m D^n$, 8
marginal conditions, 64
McCoy, 9, 28, 30
monomial from the kernel, 51
monomial rule, 50
Moyal bracket, 43

normal ordering, 82
notation, 3

one parameter families, 81

operator algebra
$e^{\xi H} A e^{-\xi H}$, 12
$e^{i\theta X + i\tau D}$, 10
adjoint, 6
adjoint of $e^{i\theta X + i\tau D}$, 11
characteristic function operator, 10
delta function, 17
differentiation, 8
exponential operator, 8
function of operators, 16
functions of operators, 7
generalized characteristic function operator, 12
Hermitian operator, 6
inverse, 6
manipulating operators, 8
phase-space operator formulas, 13
product of two operators, 7
rearrangement of operators, 22
repeated commutator, 13
representation of functions, 14
simplifying functions of operators, 20
translation operator, 9
unitary operator, 7
operator to symbol, 50, 70
ordering rule, 3

path integral approach, 91–94
configuration space, 91
phase-space, 93
phase of a signal, 100
phase-space, 3, 13, 42, 61, 62, 91, 93, 103, 139
phase-space operator formulas, 13
positive distributions, 66
probability distribution corresponding to an operator, 111
probability interpretation, 20
product of operators, 54

quantum mechanics, 98

Born interpretation, 114
quasi-distributions, 3

rearrangement of operators, 22
rearrangement operator, 70
repeated commutator, 13
representation of functions, 14
rules of association, 1, 3

scale operator, 117, 143, 148
Schwarz inequality, 123
signal, 3
singular kernels, 67
special cases, 69–84
spectrogram, 78
standard deviation of an operator, 121
standard ordering, 2, 70
star notation, 45
state function, 1, 3
symbol, 3
symmetrization ordering, 73

Taylor series association, 56, 57
terminology, 1–4
time-frequency, 95–101
    association rules, 96
    distributions, 98
    operators, 95
time-frequency space-(spatial) frequency, 100
transformation matrix, 146
transformation of differential equations into phase space, 103–106
translation invariance, 53
translation operator, 9, 89

uncertainty principle, 121–128
unitary operator, 86
unitary transformation, 85–90
    generalized association, 87
    of an operator, 86
    of the state function, 86

Weyl association, *see* Weyl operator
Weyl correspondence, *see* Weyl operator
Weyl operator, 25–46
    adjoint, 36
    algebra, 37
    commutators, 43
    conjugate of the symbol, 39
    derivative of a function, 40
    derivative of the symbol, 39
    Fourier domain, 27
    Hermiticity, 36
    inversion, 32
    linearity, 37
    Moyal bracket, 43
    operation on a function, 26
    operational form, 27
    product of operators, 43
    rearrangement procedure, 27, 30
    scaled symbol, 38
    translated symbol, 38
    unit association, 38
    Wigner distribution, 41
Wigner distribution, 3, 41, 42, 61, 65, 76, 82, 100, 103, 104, 142

ZAM ordering, 76, 78
Zassenhaus formula, 136

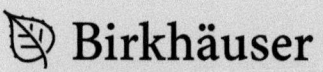 www.birkhauser-science.com

Pseudo-Differential Operators (PDO)
Theory and Applications

This series is devoted to the publication of current research in operator theory, with particular emphasis on applications to classical analysis and the theory of integral equations, as well as to numerical analysis, mathematical physics and mathematical methods in electrical engineering.

Edited by
M. W. Wong, York University, Canada
In cooperation with an international editorial board

■ **PDO 8: Unterberger, A.**, Pseudodifferential Analysis, Automorphic Distributions in the Plane and Modular Forms (2011).
ISBN 978-3-0348-0165-2

Pseudodifferential analysis, introduced in this book in a way adapted to the needs of number theorists, relates automorphic function theory in the hyperbolic half-plane Π to automorphic distribution theory in the plane. Spectral-theoretic questions are discussed in one or the other environment: in the latter one, the problem of decomposing automorphic functions in Π according to the spectral decomposition of the modular Laplacian gives way to the simpler one of decomposing automorphic distributions in $\mathbf{R}^2$ into homogeneous components. The Poincaré summation process, which consists in building automorphic distributions as series of g-transforms, for $g \in SL(2;\mathbf{Z})$, of some initial function, say in $S(\mathbf{R}^2)$, is analyzed in detail. On Π, a large class of new automorphic functions or measures is built in the same way: one of its features lies in an interpretation, as a spectral density, of the restriction of the zeta function to any line within the critical strip.

■ **PDO 7: de Gosson, M.**, Symplectic Methods in Harmonic Analysis and in Mathematical Physics (2011).
ISBN 978-3-7643-9991-7

The aim of this book is to give a rigorous and complete treatment of various topics from harmonic analysis with a strong emphasis on symplectic invariance properties, which are often ignored or underestimated in the time-frequency literature. The topics that are addressed include (but are not limited to) the theory of the Wigner transform, the uncertainty principle (from the point of view of symplectic topology), Weyl calculus and its symplectic covariance, Shubin's global theory of pseudo-differential operators, and Feichtinger's theory of modulation spaces. Several applications to time-frequency analysis and quantum mechanics are given, many of them concurrent with ongoing research.

This book is primarily directed towards students or researchers in harmonic analysis (in the broad sense) and towards mathematical physicists working in quantum mechanics. It can also be read with profit by researchers in time-frequency analysis, providing a valuable complement to the existing literature on the topic. A certain familiarity with Fourier analysis and introductory functional analysis (e.g. the elementary theory of distributions) is assumed. Otherwise, the book is largely self-contained and includes an extensive list of references.

■ **PDO 6: Gupur, G.**, Functional Analysis Methods for Reliability Models (2011).
ISBN 978-3-0348-0100-3

The main goal of this book is to introduce readers to functional analysis methods, in particular, time dependent analysis, for reliability models. Understanding the concept of reliability is of key importance – schedule delays, inconvenience, customer dissatisfaction, and loss of prestige and even weakening of national security are common examples of results that are caused by unreliability of systems and individuals.
Functional Analysis Methods for Reliability Models is an excellent reference for graduate students and researchers in operations research, applied mathematics and systems engineering.

■ **PDO 5: Wong, M.W.**, Discrete Fourier Analysis (2011).
ISBN 978-3-0348-0115-7

This textbook presents basic notions and techniques of Fourier analysis in discrete settings. Written in a concise style, it is interlaced with remarks, discussions and motivations from signal analysis.
The book is aimed at advanced undergraduate and graduate students in mathematics and applied mathematics. Enhanced with exercises, it will be an excellent resource for the classroom as well as for self-study.

MIX
Papier aus verantwortungsvollen Quellen
Paper from responsible sources
FSC® C105338

If you have any concerns about our products,
you can contact us on
ProductSafety@springernature.com

In case Publisher is established outside the EU,
the EU authorized representative is:
**Springer Nature Customer Service Center GmbH
Europaplatz 3, 69115 Heidelberg, Germany**

Printed by Libri Plureos GmbH
in Hamburg, Germany